FIBER OPTIC INSTALLER'S FIELD MANUAL

Other McGraw-Hill Books of Interest

FIBER OPTIC INSTALLER'S FIELD MANUAL

Bob Chomycz

McGraw-Hill

New York   San Francisco   Washington, D.C.   Auckland   Bogotá
Caracas   Lisbon   London   Madrid   Mexico City   Milan
Montreal   New Delhi   San Juan   Singapore
Sydney   Tokyo   Toronto

Library of Congress Cataloging-in-Publication Data

Chomycz, Bob.
    Fiber optic installer's field manual   /   Bob Chomycz.
       p.   cm.
    ISBN 0-07-135604-5
    1.

  00000.0000   2000
  000.0'000—dc21   00-00000

Trademarks
Kevlar is a registered trademark of DuPont.
ST is a registered trademark of AT&T, USA.

McGraw-Hill

A Division of The McGraw-Hill Companies

1 2 3 4 5 6 7 8 9 0 DOC/DOC 9 0 9 8 7 6 5 4 3 2 1 0 9

ISBN 0-07-135604-5

The sponsoring editor of this book was Steve Chapman, the editing supervisor was Sally Glover, and the production supervisor was Pamela Pelton. It was set in the HB1 design in Times Roman by Joanne Morbit and Michele Pridmore of McGraw-Hill's Professional Book Group composition unit.

Printed and bound by R. R. Donnelley & Sons Company.

McGraw-Hill books are available at special quantity discounts to use as premiums and sales promotions, or for use in corporate training programs. For more information, please write to the Director of Special Sales, Professional Publishing, McGraw-Hill, Two Penn Plaza, New York, NY 10121-2298. Or contact your local bookstore.

This book was printed on recycled, acid-free paper containing a minimum of 50% recycled, de-inked fiber.

CONTENTS

Chapter 16. Lightwave Equipment   155

Chapter 17. WDMS and Other Optical Components   165

Chapter 18. SONET/SDH   183

Chapter 19. LAN   195

Chapter 20. Fiber System Deployment   201

PREFACE

I am very pleased to offer to you *Fiber Optic Installer's Field Manual*. The primary objective of this book is to introduce the reader to fiber optic communications, fiber components, and installation techniques. My emphasis is on practical installation techniques using industry standards.

The approach I have chosen for this book is not highly theoretical, nor do I attempt to cover the physics of lightwave communications. There are already many good books available that discuss the theoretical aspects of fiber optics, but only a few that deal with the practical aspects of fiber optics in the real world. Because of the unique nature of this medium, many conventional electrical wire implementation techniques are not applicable. Employees working with fiber optics need to understand not only the basic theory but also the practical methods used to implement this technology. This book attempts to accomplish that task.

Fiber Optic Installer's Field Manual will be of primary interest to the following professions: technicians, electricians, cable installers, testers, telecommunications personnel, data communications personnel, engineers, technologists, tradespeople, students, and any others with an interest in the field, regardless of their level of knowledge.

Bob Chomycz

CHAPTER 1
INTRODUCTION

1. 1 THE FIBER OPTIC REVOLUTION

Over the last 20 years, a quiet revolution has been changing the world of communications. Indirectly, it will affect all our lives by enhancing our ability to communicate large amounts of information over vast distances with extreme clarity and reliability. This revolution centers on the replacement of existing copper wire communication cables with thin strands of glass fibers that carry light pulses.

Since the earliest recorded history, light has been used to communicate over distances, although the techniques employed have often been slow and cumbersome. Communication has been limited by atmospheric conditions—usually an impossibility in fog or heavy rain, for example—and restricted to line-of-site operation. As early as the ancient Greeks and Phoenicians, sunlight was reflected off mirrors for signaling between towers, and this technique continued into the modern era with several variations. Eventually, sunlight was replaced by artificial light, and the on/off signaling became more structured until it resembled a Morse code. The military, for example, still uses a version of this technique for low-speed communication between ships.

In the late nineteenth century, Alexander Graham Bell worked on a design for a "Photophone," for sending voice over a light beam. Sunlight was reflected off a mirror that vibrated to voice sound waves. The receiver was a photocell connected to an electric current that passed to a speaker. The idea was good, but the technology was not yet in place to use it practically.

After the laser was invented in 1958, further studies were performed with light communication in air. Lasers provided a narrow band of light radiation that could be bent with mirrors. Communication by light was still not practical because it required a clear line-of-sight; fog or rain still presented a problem of obstruction to the link. Experiments continued with light propagation in a glass medium. Conclusions showed glass to be preferable to air for two reasons—its constant nature and its ability to remain unaffected by environmental variations.

In 1970, the first low-loss optical fiber was developed. The optical fiber, made from silica glass about 250 micrometers (mm) in diameter—about the size of a human hair—was used to propagate light in a lab environment. This was the beginning of fiber optics.

Shortly after this, the process of manufacturing thin glass fiber strands was perfected and, in the mid 1970s, Corning Inc. made fiber optic cable available commercially. This launched the fiber optic revolution. Short-distance systems were subsequently tested by many telephone companies. With the continual refinement of the technology, communication distances increased and more products became available.

In 1980, Bell announced the installation of 611 miles of optical fiber in its northeast U.S. corridor. Likewise, Saskatchewan Telephone announced the installation of 3600 km of optical fiber in Canada. At the 1980 Lake Placid Winter Olympics, fiber optics was first used to transmit television signals. In the years that followed, fiber optics gradually gained popularity in the telecommunications world. Today, it is a widely accepted and proven technology. The replacement of old wire communication lines with new fiber optic cable is now the norm for many applications.

The basic principle of the on/off light communication used in the past is similar to the principle used in fiber optics today. The information signal to be transmitted controls a light source by turning it on and off in a particular coded sequence, or by varying its intensity. The light is then coupled into an optical fiber that guides it for the distance of the communication. At the receiving end, a detector decodes the light and reproduces the information signal.

Although light travels in a straight line in free space, the glass properties of the optical fiber guide the light around bends and allow fiber optic cable routes to work like standard copper wire cable routes, with some restrictions. The distance of propagation is determined mainly by the loss of the light in the optical fiber and by the rate of the on/off signaling. At the other end of the optical fiber the light is coupled to a light-sensitive photodetector which converts the pulsing light signal back to a usable electrical signal.

Today's fiber optic technology can support transmission rates of over 9 billion light pulses per second. This translates to over 129,000 simultaneous telephone calls. A standard 200-fiber cable can carry over 12 million telephone conversations compared to the 10,000 conversations a similar-sized copper cable can carry.

Optical fiber doesn't care what type of signal it is propagating. This makes it a versatile medium, available for practically all types of communication, including telephone, video, television, images, computers, local area networks (LANs), wide area networks (WANs), control systems, and so on. One fiber optic cable installation can be used for many applications.

As this lightwave revolution continues to develop and grow, we can expect better and more abundant service for our everyday needs. In fact, we are already benefiting from optical fiber's versatility in many practical areas of our lives. Already, many of our phone calls use fiber optic facilities. Cable TV companies are adding fiber optics to their networks so we can enjoy a larger selection of channels with better quality. Interactive television, which uses fiber optics, is being tested in many locations. Fiber optics is also used in many industries to carry high-speed computer data throughout a factory or across countries.

1.2 BASIC TRANSMISSION

Fiber optics involves the transmission of information by light through long transparent fibers made from glass or plastic. A light source modulates a light-emitting diode (LED), or a laser turns on or off or varies in intensity in a manner that represents the electrical information input signal. The modulating light is then coupled to an optical fiber that propagates the light. An optical detector at the opposite end of the fiber receives the modulating light and converts it back to an electrical signal which is identical to the input signal.

Light transmission techniques can be divided into three major categories: digital modulation, analog modulation, and digital modulation with analog-to-digital conversion. Digital modulation involves the conversion of the electrical digital input signal into a similar coded sequence of on or off (digital) light pulses (see Fig. 1.1a). Because all computer communications use electrical digital communications, this type of modulation is well suited for computer data transmission.

Analog communication signals, such as for voice or video transmission, vary in electrical amplitude and period. Analog modulation converts this electrical input signal into an optical signal of similarly varying light intensity (see Fig. 1.1b). This technique can be relatively inexpensive and is often used in fiber optic modem applications.

Analog signals can also be converted to a digital format using an analog-to-digital converter (A-to-D converter) before the modulation step. Digital light signals are then propagated in the optical fiber (see Fig. 1.1c). At the other end, the digital light signal is converted to an electrical digital signal by the detector. Then a second analog-to-digital converter converts the digital signal back into its original analog form. This technique provides signal conformity with other digital signals and allows a number of signals to be combined into an optical fiber aggregate using multiplexing equipment.

The transmission techniques shown in Fig. 1.1 show information transmission in only one direction. However, most systems require full, simultaneous two-way communications. Therefore, a second identical set of modulation and detection devices is implemented in the opposite direction to form a fully functional two-way communication system (see Fig. 1.2).

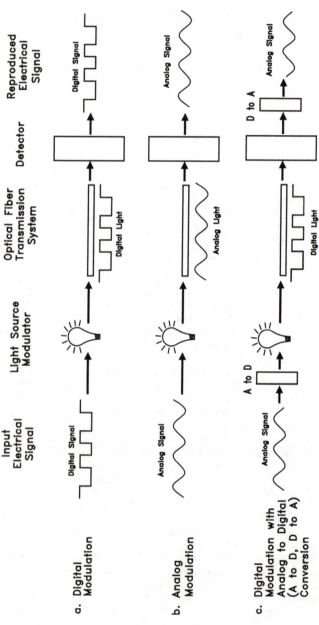

FIGURE 1.1 Fiber optic transmission basics.

4

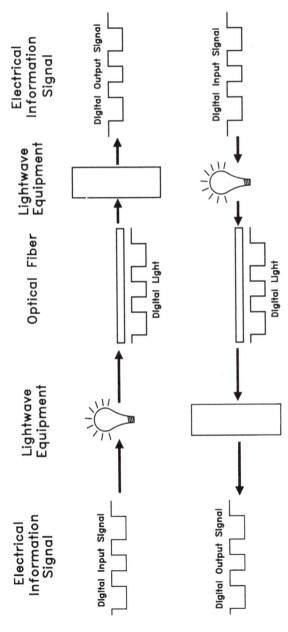

FIGURE 1.2 Two-way communications.

1.3   ADVANTAGES AND DISADVANTAGES

Optical fiber has become a popular medium for many communications requirements. Its appeal can be attributed to the many advantages that optical fiber has over conventional electrical transmission methods This lightwave transmission medium also has drawbacks, however, which should be examined before proceeding with an installation. The following sections describe some of these considerations.

Advantages

Large Capacity.   Optical fiber has the capacity to transmit large amounts of information. With present technology, over 2,000,000 simultaneous telephone conversations can be placed on only two optical fibers. A fiber optic cable can contain as many as 200 optical fibers, which would increase the link capacity to over 200,000,000 conversations. Compare this to conventional wire facilities, in which a large multipair cable can carry 500 conversations, a coaxial cable can carry 10,000 conversations, and a microwave radio or satellite link can carry 2000 conversations.

Size and Weight.   Fiber optic cable is much smaller in diameter and lighter in weight than a copper cable of similar capacity. This makes it easier to install, especially in existing cable locations (such as building risers) where space is at a premium.

Electrical Interference.   Optical fiber is not affected by electromagnetic interference (EMI) or radio frequency interference (RFI), and it does not generate any of its own interference. It can provide a clean communication path in the most hostile EMI environment. Electrical utilities use optical fiber along high-voltage lines to provide clear communication between their switching stations. Optical fiber is also free of cross talk between fibers. Even if light is radiated by one optical fiber, it cannot be recaptured by another optical fiber.

Insulation.   Optical fiber is an insulator. The glass fiber eliminates the need for electric currents for the communication path. Proper all-dielectric fiber optic cable contains no electrical conductor and can provide total electrical isolation for many applications. It can eliminate interference caused by ground loop currents or potentially hazardous conditions caused by electrical discharge onto communication lines such as lightning or electrical faults. It is an intrinsically safe medium often used where electrical isolation is essential.

Security.   Optical fiber offers a high degree of security. An optical fiber cannot be tapped by conventional electrical means such as surface conduction or electromagnetic induction, and it is very difficult to tap onto optically. Light rays travel down the center of the fiber and few or none of them escape. Even if a tap is successful, it can be detected by monitoring the optical power received at the termination. Radio or satellite communication signals can easily be captured for decoding.

Reliability and Maintenance.   Optical fiber is a constant medium and is not subject to fading. Properly designed fiber optic links are immune to adverse temperature and moisture conditions and can even be used for underwater cable. Optical fiber also has a long service life span, estimated at over 30 years for some cables. The maintenance required for a fiber optic cable is minimal; there is no copper in the cable that can corrode and cause intermittent or lost signals; and the cable is not affected by short circuits, power surges, or static electricity.

Versatility. Fiber optic communications systems are available for most data, voice, and video communications formats. Systems are available for RS232, RS422, V.35, Ethernet, Arcnet, FDDI, T1, T2, T3, Sonet, 2/4 wire voice, E&M signal, composite video, and many more.

Expansion. Properly designed fiber optic systems can easily be expanded. A system designed for a low data rate, for example, T1 (1.544 Mbps), can be upgraded to a higher data rate system, OC-12 (622 Mbps), by changing the electronics. The fiber optic cable facility can remain the same.

Signal Regeneration. Present technology can provide fiber optic communication beyond 70 km (43 mi) before signal regeneration is required, which can be extended to 150 km (93 mi) using laser amplifiers. Future technology may extend this distance to 200 km (124 mi) and possibly 1000 km (621 mi). The savings in intermediate repeater equipment costs and in maintenance can be substantial. Conventional electrical cable systems, by contrast, can require repeaters every few kilometers.

Disadvantages

Electrical-to-Optical Conversion. Before connecting an electrical communication signal to an optical fiber, the signal must be converted to the lightwave spectrum [850, 1310, or 1550 nanometers (nm)]. This is performed by the electronics at the transmitting end, which properly formats the communication signal and converts it to an optical signal using an LED or solid-state laser. This optical signal is then propagated by the optical fiber. At the receiving end of the optical fiber, the optical signal must be converted back to an electrical signal to be useful. The cost of this additional electronics needed to convert the signal to light and back to an electrical signal should be considered in all applications.

Right of Way. A physical right of way for the fiber optic cable is required. The cable can be directly buried, placed in ducts, or strung aerially along the right of way. This may require the purchase or leasing of property. Some rights of way may be impossible to acquire. For locations such as mountainous terrain or some urban environments, other wireless communication methods may be more suitable.

Special Installation. Because optical fiber is predominantly silica glass, special techniques are needed for the engineering and installation of the fiber cable. Conventional wire cable installation methods, crimping, wire wrapping, or soldering, for example, no longer apply. Proper fiber optic equipment is also required to test and commission the optical fibers. Technicians must be trained in the installation and commissioning of the fiber optic cable.

Repairs. Fiber optic cable that becomes damaged is not as easily repaired as many copper cables. Repair procedures require a skilled technical crew with proper equipment. In some situations, the entire cable may need to be replaced. This problem can be further complicated if a large number of users rely on the facility. A proper system design with physically diverse routing to accommodate such contingencies is thus important.

Although there may be many advantages that favor a fiber optic installation, they should be weighed carefully against the disadvantages of each application. All the costs of the implementation and operation of fiber optic facilities should be analyzed.

1.4 APPLICATIONS

Fiber optic cable is currently being used as a communication medium for many different applications. Many telephone companies, for example, are deploying fiber optics to provide communication between their central offices (COs), throughout cities, across countries, and over long oceanic routes (see Fig. 1.3). Plans now exist to extend fiber right into the home for high-quality video telephone transmissions. Fiber optics provides a reliable, high-capacity link for voice, data, and video traffic.

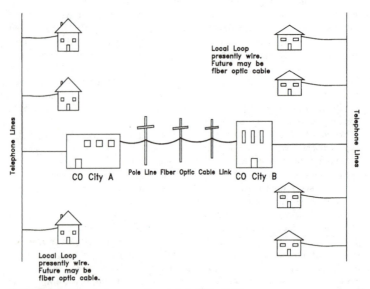

FIGURE 1.3 Fiber optics connecting central offices (COs).

Cable television companies are deploying fiber optic cable to carry high-quality signals from their head end center to hub locations distributed around cities (see Fig. 1.4). Fiber optics improves the quality of television signals and increases the number of available channels. Future plans may involve connecting optical fiber directly into the home to provide many new services for the user. Such services as interactive television, banking at home, or working from a home office system are planned for the future.

Fiber optics is ideal for data communications. Very high data rates can be achieved on a thin fiber optic cable. Signals are not distorted by office interference and do not cause any interference of their own. The dielectric properties of fiber optics provide a safe interface between computers, terminals, and workstations. There is no chance of harmful ground current loops endangering users or damaging expensive computing equipment.

Many computer centers are using fiber optics to provide high-speed data communications in their LANs (see Fig. 1.5). Many products are available for many different applications. High-speed data highways, such as fiber distributed data interface (FDDI), asynchronous transfer mode (ATM), Gigabit Ethernet, and SONET, are available to provide backbone connectivity with various networks. These new technologies offer such benefits as high data transmission rates, increased distances, and reliable and secure communications. Large quantities of data can now quickly and efficiently traverse large geographical areas.

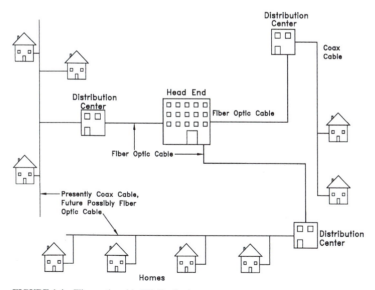

FIGURE 1.4 Fiber optic cable TV distribution.

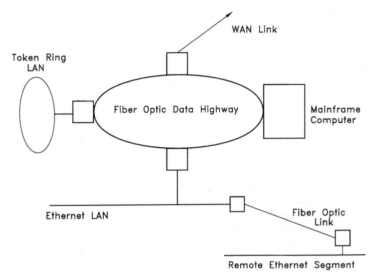

FIGURE 1.5   Fiber optics in a local area network (LAN) environment.

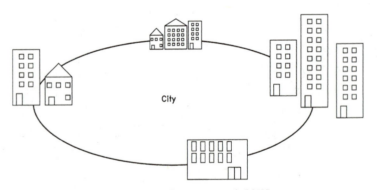

FIGURE 1.6   Fiber optics in a metropolitan area network (MAN).

Because fiber optics is a cost-effective investment, companies are installing it in metropolitan areas. These metropolitan area networks (MANs) meet all present communications requirements and allow for ample expansion of future systems (see Fig. 1.6).

Businesses can be geographically distributed but remain connected and on line at all times. The benefits can be enormous. Banks, for example, can keep all branches connected at high data rates for all types of transactions.

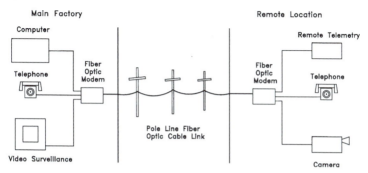

FIGURE 1.7   Fiber optics in an industrial environment.

Commercial outlets can maintain better centralized control of their point-of-sale computer systems. Inventory and pricing adjustments can be made on demand. Large firms with extensive computing requirements, such as brokerage houses, insurance companies, and government offices, can maintain secure, continuous, high-data-rate communications. Industries can keep close watch on the production and inventory of their distributed plants.

Industry is using fiber optic communication to improve the reliability and capacity of data and control transmissions. Because of the inherent nature of light communication, it is immune to all electrical interference caused by large motors, switches, lights, and other mechanisms commonly found in industrial environments. Optical fiber's versatility allows computer data, telephone, video, control, and sensor transmissions all to be placed onto one fiber optic cable (see Fig. 1.7).

CHAPTER 2
PROPERTIES OF LIGHT

2.1 ELECTROMAGNETIC SPECTRUM

Light behaves as an electromagnetic wave and belongs to the electromagnetic spectrum (EMS). The number of oscillations per second that an electromagnetic wave completes is called its frequency. Visible light has a frequency of around 2.3×10^{14} cycles per second. One cycle per second is more commonly referred to as hertz (Hz). As Fig. 2.1 shows, light frequencies are much higher than other electromagnetic wave frequencies, such as radio waves and television waves.

Wave Theory

$\lambda = c/f$

where λ = wavelength in meters
c = the speed of light in the medium
f = the frequency of light in cycles per second or hertz

The electromagnetic wavelength (λ) is the length in meters of one cycle of a wave. Visible light wavelength is in the range of 770×10^{-9} to 330×10^{-9} m. One billionth of a meter (10^{-9} m) is commonly referred to as a nanometer (nm). A mathematical relationship between frequency and wavelength exists as $\lambda = c/\text{frequency}$, where c is the speed of the light in the material where it propagates. Light is normally referred to by its wavelength value in nanometers rather than its frequency.

Light used for fiber optic communication exists in the infrared (IR) region of the spectrum, just below visible light. The windows of the fiber optic communication spectrum are in the 1550 nm, 1310 nm, and 850 nm

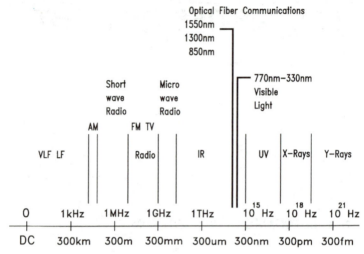

FIGURE 2.1   Electromagnetic spectrum.

bands. Light visible to the human eye begins at about 770 nm (red) and ends at 330 nm (blue). Therefore, optical fiber light is generally not visible to the eye. In some circumstances, when broad-spectrum light-emitting diodes (LEDs) are used for 850-nm transmission, a portion of the LED's spectrum may fall into the visible range. A deep red light may then be visible.

Extreme caution must be exercised around fiber optic light. Whether visible or invisible, certain forms of fiber optic light can cause damage to the eye. Care should be exercised at all times to ensure that fiber optic light does not enter the eye either directly or from indirect surface reflections (see Chap. 6 on safety precautions).

The high-frequency nature of fiber optic light ($\sim 2 \times 10^{14}$ Hz) allows the light to carry information at very high data rates. Present-day transmission equipment can modulate the light at 9 gigabits per second (9×10^9 bps). This is much higher than conventional electrical transmission media and is not a maximum data rate.

Light also behaves as a particle called a photon and has energy E that can be defined using the formula $E = hc/\lambda$ and expressed in units called joules, where h is Planck's constant, equal to 6.626×10^{-34} J, λ is the wavelength of light in meters, and c is the speed of light in the propagating material (in space, $c = 2.998 \times 10^8$ m/s). Power is defined as the rate at which energy is delivered; therefore, power in watts can be written as $P = E/t$, where t is time. The particle behavior of light explains how sources generate light and how detectors are able to convert light back to electrical energy.

Particle Theory

$E = hc/\lambda$

where E = photon's constant energy in joules
h = Planck's constant, 6.626×10^{-34} J
c = the speed of light in the medium
λ = the wavelength of light in meters

2.2 LIGHT PROPAGATION

In free space, light travels in straight lines at a speed of 299,800 km/s or 186,292 mi/s. The direction that light waves travel is called a light ray and is used in fiber optics to explain many fiber characteristics.

When a light ray passes from one material to another different material, it changes speed and direction at the material's boundary. If the second material is transparent, some of the light enters the material. At the boundary

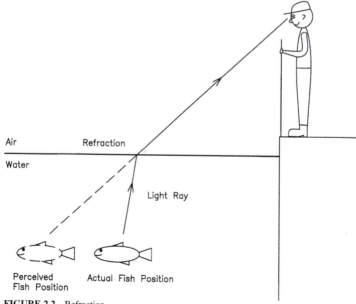

FIGURE 2.2 Refraction.

between the two materials, the light ray bends before continuing in the second material. This bending is called refraction.

For example, if a person is standing at the edge of a lake and sees a fish in the water, the fish's physical location is different from the location it appears to have to the observer. As a light ray travels from the fish in the water (first material) to the person standing in the air above the surface (second material) it encounters the water-air boundary. At this boundary, the light ray is refracted (bent) before continuing in a straight line to the person (see Fig. 2.2).

Because of the light ray's refraction, the fish's actual physical location is different from the position it appears to occupy. If the person wanted to catch the fish, he or she would therefore need to aim slightly off the fish's perceived position. In addition, when light passes from one material to another different material, some of the light does not enter the second material but is reflected back into the first material.

For example, a person watching a sunset over a lake can see the sun's reflection in the lake. Light rays emitted by the sun travel directly to the person, while other sunlight rays strike the water's surface. Light rays that strike the water's surface are either reflected back into the atmosphere and seen by the observing person or refracted into the water (and seen by the fish). Light is reflected at the water-air boundary, but not completely (see Fig. 2.3).

This same behavior of light would occur if the light source were under water. Some of the light striking the water-air boundary would be reflected back into the water, and the rest would be refracted into the air. An addi-

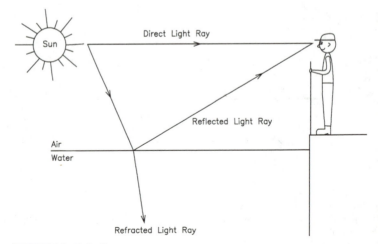

FIGURE 2.3   Reflection.

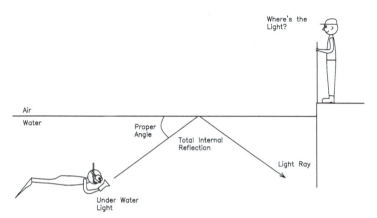

FIGURE 2.4 Total reflection.

tional phenomenon occurs in this situation. At the proper angle, all the light rays that strike the water-air boundary are reflected back into the water—none escape into the air (see Fig. 2.4). This phenomenon, also known as total reflection, is the basis for the confinement of light in an optical fiber (multimode fiber only).

The water is analogous to a fiber core, and the air to the fiber cladding. Light launched in the core totally reflects at the core cladding boundary and propagates in the core.

CHAPTER 3
OPTICAL FIBER

3.1 OPTICAL FIBER COMPOSITION

An optical fiber is a long, cylindrical, transparent material that confines and propagates light waves (see Fig. 3.1). It is comprised of three layers: the center core that carries the light, the cladding layer that covers the core which confines the light to the core, and the coating that provides protection for the cladding.

The core and cladding are commonly made from pure silica glass (SiO_2), while the coating is a plastic or acrylate cover. It is interesting to note just how transparent the glass is in the core of an optical fiber. The picture window you have at home in your living room is $1/8$-in-thick plate glass. You could replace it with a window made of fiber optic core glass that was 3 mi thick, and you would get the same bright image coming through the 3-mi-thick window that you currently do with the $1/8$-in plate glass window.

The core and cladding layers differ slightly from each other in their composition due to small quantities of material such as boron or germanium added during the manufacturing process. This alters the index of refraction characteristic of both layers, resulting in the light confinement properties needed to propagate the light rays.

The index of refraction in the silica core is approximately 1.5 and in the cladding is slightly less, about 1.48. The index of refraction of air is approximately 1.0. The fiber coating is colored using manufacturer's standard color codes to facilitate the identification of fiber. Like glass, optical fibers can be made completely from plastic. Optical fibers which are made from plastic are usually less expensive but have higher attenuation (loss) and limited application.

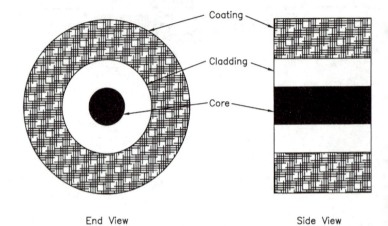

End View   Side View

FIGURE 3.1   Optical fiber.

Common Fiber Diameters.

Optical fibers used for telecommunications are manufactured in five major core and cladding diameters, as illustrated in the following table:

Common Optical Fiber and Buffer Diameters (μm)

	Core	Cladding	Coating	Buffer or tube
I	7 to 10	125	250 or 500	900 or 2000–3000
II	50	125	250 or 500	900 or 2000–3000
III	62.5	125	250 or 500	900 or 2000–3000
IV	85	125	250 or 500	900 or 2000–3000
V	100	140	250 or 500	900 or 2000–3000

Note: 250 μm = 0.25 mm (millimeters).

Fiber size is specified in the format "core/cladding." Therefore, a 62.5/125 fiber means the fiber has a core diameter of 62.5 μm and cladding diameter of 125 μm.

The coating covers the cladding and can be either 250 or 500 μm in diameter. For a tight-buffered cable construction, a 900-μm-diameter plastic buffer covers the coating. For a loose tube cable construction, the fiber, with a 250-μm coating, lies loose in a 2- to 3-mm plastic tube (see Chap. 4).

I. Core: 7 to 10 μm.   A fiber with a core diameter of 7 to 10 μm is referred to as a single-mode fiber. It can propagate the highest data rate. It is commonly used for long-distance (over 2 km) or high-data-rate transmission applications. Due to the small diameter of its core, lightwave equipment needs to employ high-precision connectors and laser light sources in order to couple light in and out of the fiber. This inflates equipment prices. Single-mode fiber optic equipment is often priced much higher than multimode equipment. However, single-mode fiber optic cable is generally less expensive than multimode cable.

II. Core: 50 μm.   The 50-μm core diameter fiber was the first telecommunications fiber to sell in large quantities and is still in common use today. Its low numerical aperture (NA, see Sec. 3.2) and smallest multimode core size result in the least amount of LED source light being coupled into the fiber. Of all the multimode fibers, this fiber has the highest potential bandwidth and is gaining popularity due to applications that demand high bandwidth such as Gigabit Ethernet.

III. Core: 62.5 μm.   The 62.5-μm core diameter fiber is popular for multimode transmission and was the standard for many applications. This fiber has a lower potential bandwidth than 50/125 fiber. Its higher NA and core diameter provide slightly better LED light-coupling power into it than the 50/125-μm fiber. This makes a longer distance possible. However, because of its lower bandwidth, the 62.5-μm fiber is losing ground to the 50/125-μm fiber.

IV. Core: 85 μm.   The 85/125-μm fiber is a European fiber size and is not popular in North America. It has good light-coupling ability, similar to the 100-μm core diameter, and uses the standard 125-μm diameter cladding. This allows the use of standard 125-μm connectors and splices with this fiber.

V. Core: 100/140 μm.   The 100/140-μm multimode fiber's large core diameter makes it the easiest fiber to connect. It is less sensitive to connector tolerances and the accumulation of dirt on connectors. It couples the most light from the source but has a significantly lower potential bandwidth than other smaller core sizes. It can be found in intermediate-length, connector-intensive spans (in buildings) that have low-data-rate requirements. Since it is not common, it may be difficult to obtain.

There are other larger core diameters, but they are even less common and their applications are limited. They are used primarily for short connection spans (between equipment) or in applications other than data communications such as visual light transmission.

The following table summarizes these fiber core sizes and their characteristics:

Optical Fiber Characteristics

	Core	NA	Loss	Bandwidth	Wavelength
I	7 to 10	Smallest	Lowest	Highest	1310 or 1550
II	50	Smaller	Lower	Higher	850 or 1310
III	62.5	Medium	Low	Medium	850 or 1310
IV	85	Large	High	Lower	850 or 1310
V	100	Largest	Higher	Lowest	850 or 1310

Matching Optical Fibers.   When matching multimode fibers for splicing or a connection, the core diameters should be the same size. A 62.5/125-μm multimode fiber should only be spliced to another multimode 62.5/125-μm fiber. If necessary, two multimode fiber sizes can be mixed. A 62.5/125-μm fiber can be connected to a 50/125-μm fiber. However, because of the mismatch in core diameters, a large loss at the connection would result. The loss is proportional to the cores of the two fibers' cross-sectional area ratio. A 62.5/125-μm fiber connecting to a 50/125-μm fiber would result in a mismatch loss of at least 2 dB. The connection loss would be the 2-dB mismatch loss plus the connector loss.

Single-mode fibers should not be spliced or connected to multimode fibers. Due to different principles of propagation in the two fiber types, the resulting light transmission would be either poor or none at all. When splicing single-mode fibers, the mode field diameter is used to match the fibers instead of the core diameter. The closer the similarity to the mode field diameters, the lower the splice loss.

Equipment designed for single-mode fiber should only be connected to single-mode fibers. Equipment designed for multimode fiber should only be connected to multimode fibers.

Fiber optic equipment specifications should be carefully reviewed to determine the proper fiber type and core diameter. One fiber core diameter can be specified to be used with the lightwave equipment, or a number of different sizes can be listed for the equipment (for multimode fiber).

Because of the higher light-coupling power of larger core diameter fibers, longer transmission distances can be achieved for some applications when larger core diameters are used.

The following multimode fiber table is an example that illustrates how cable length increases with an increase in core diameter (though only for some lightwave equipment). Even though optical fiber attenuation increases with an increase in core diameter, more light is coupled into the larger core diameter, and the fiber transmission length increases. Similar charts are available for selected lightwave equipment.

Fiber size (μm)	Fiber attenuation (dB/km)	NA	Cable length (km)
50/125	4.0	0.20	0.2
50/125	3.0	0.20	0.27
50/125	2.7	0.20	0.3
62.5/125	4.0	0.29	1.3
62.5/125	3.7	0.29	1.5
100/140	5	0.29	1.5
100/140	4	0.29	1.8

Single-mode fiber manufacturers have also designed special fibers with larger cores. The effect is the same. The larger core enables more light to be captured by the fiber, resulting in further transmission distance.

3.2 LIGHT TRANSMISSION IN A FIBER

When a light ray travels unimpeded through a medium, such as air or glass, it travels in a straight line. However, when a light ray passes from one medium to another, it bends at the medium boundary. This bending is called refraction. The angle at which it is refracted is called the angle of refraction. The angle at which the light ray strikes the medium boundary is called the angle of incidence.

The angle of incidence is mathematically related to the angle of refraction according to Snell's law.

Snell's Law

$$n1 \times \sin a = n2 \times \sin b$$

where $n1$ = the index of refraction of the first material
$n2$ = the index of refraction of the second material
a = the angle of incidence in the first material
b = the angle of refraction in the second material

Light ray refraction occurs at the end of a fiber where it passes between air and the fiber core medium. The angles of refraction and incidence are measured to the axis perpendicular to the air fiber boundary. For a properly cleaved fiber, this axis is the same as the fiber's axis (see Fig. 3.2).

Optical fiber cleaving is the process of cutting an optical fiber in such a manner that a smooth, flat end surface perpendicular to the fiber's axis is produced. This will ensure the propagation of the maximum amount of light in the fiber.

Only light rays that are incident on the air-fiber boundary at angles less than the maximum coupling angle are refracted into the fiber core and captured by the fiber (see Fig. 3.3).

The fiber's numerical aperture (NA) is mathematically related to the maximum coupling angle.

Numerical Aperture (NA)

$$NA = \sin (\text{maximum coupling angle})$$

or

$$NA = (n1^2 - n2^2)^{1/2}$$

where NA = the fiber's numerical aperture
$n1$ = the index of refraction of the core
$n2$ = the index of refraction of the cladding

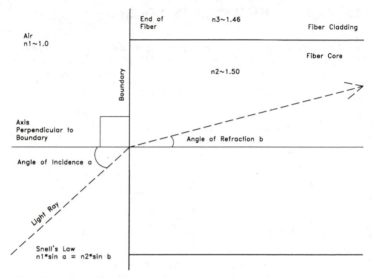

FIGURE 3.2 Refraction at an optical fiber air-fiber boundary.

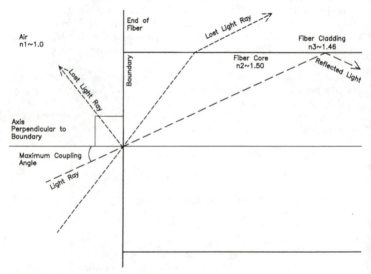

FIGURE 3.3 Maximum coupling angle.

Typical maximum coupling angles for a multimode fiber vary from 10 to 30°. Typical NA values vary from 0.2 to 0.5. The NA value is normally specified for an optical fiber.

When a light ray passes through a medium boundary from a medium with a high index of refraction to a medium with a lower index of refraction, the ray is refracted at the boundary into the second medium. As the light ray's incident angle increases, a point is reached at which the light ray is no longer refracted into the second medium and is completely reflected back into the first medium. This is called total internal reflection, and the angle at which it occurs is called the critical angle (see Fig. 3.4). The critical angle can be determined by the following formula:

Critical Angle

Critical angle = arcsin $(n2/n1)$

where $n1$ = the index of refraction of the first material
$n2$ = the index of refraction of the second material

Total internal reflection occurs when light in a high index of refraction medium strikes a boundary with a lower refractive index medium. This

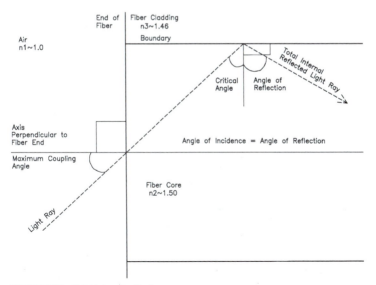

FIGURE 3.4 Total internal reflection.

phenomenon occurs at the multimode fiber's core cladding boundary and is responsible for confining the light in the fiber core.

Light rays entering the multimode fiber core at angles less than the maximum coupling angle strike the core cladding boundary at angles greater than the critical angle. They are then totally reflected back into the core and travel to the next core cladding boundary to be reflected again. If the fiber is bent, the light rays' angle of incidence decreases. Bending the fiber past the fiber's minimum bend radius will cause a significant fraction of the light rays to decrease their incident angle past the critical angle, and they will be refracted into the cladding and lost.

The fiber manufacturer will specify a minimum bending radius for a fiber cable. Bending the fiber cable tighter than this radius will result in fiber attenuation at the bend.

3.3   MULTIMODE FIBER

A multimode fiber is a fiber that propagates more than one mode of light. The maximum number of modes of light (light ray paths) that can exist in a fiber core can be estimated mathematically by the following expression:

Number of Modes in a Fiber Core

$$M = 1 + 2D (n1^2 + n2^2)^{0.5}/\lambda$$

where D = core diameter
$n1$ = the core index of refraction
$n2$ = the cladding index of refraction
λ = the wavelength of light

For multimode fiber, the number of modes can easily be over 1000. The numbers of modes that actually exist depend on other fiber characteristics and can be reduced during propagation.

Multimode fiber is commonly used in short-distance (usually less than 2 km) communication applications. Terminating electronics are less expensive and are typically simpler in design when used for multimode transmissions. A light-emitting diode (LED) is commonly used as the light source. Multimode fiber's core size is larger than single-mode fiber and is therefore easier to connect, with a greater tolerance of components with less precision.

There are two types of multimode fibers: step index fiber and graded index fiber. They differ by the index of refraction profiles of their core and cladding.

Step Index Fiber. A step index fiber is an optical fiber that has a core and cladding with different but uniform refractive indexes. At the core-cladding boundary there is an abrupt change in the refractive index. Light confinement in any step index fiber is due to the reflection property at the core-cladding boundary. This is caused by the difference in the refractive index of the two materials. As Fig. 3.5 shows, light rays are reflected at this boundary and propagated along the fiber.

The light rays travel many different paths in the fiber core. Because the distance differs that individual rays must travel, they will arrive at their destinations at different times. This results in a transmitted pulse that spreads over time.

As Fig. 3.6 shows, light rays d1, d2, and d3 begin at the same time t, but after traveling the length of the fiber they arrive at their destinations at different times because of the different propagation paths in the fiber core. As predicted, this results in the pulse spreading over time. This signal distortion is called multimode (modal) dispersion.

Such pulse spreading restricts the data transmission rate because data rate is inversely proportional to pulse width. A wider pulse means fewer pulses can be sent per second, and this results in a lower transmission bandwidth.

This is the primary data rate limiting factor in a multimode fiber:

$$\text{Data rate transmission } \alpha \text{ } 1/\text{pulse width}$$

Graded Index Fiber. The refractive index of the graded index fiber core decreases from the center outward. The refractive index of the cladding is uniform. Graded index fiber bends light rays into wavelike paths because of the core's nonuniform refractive index (see Fig. 3.7). The outer region of the

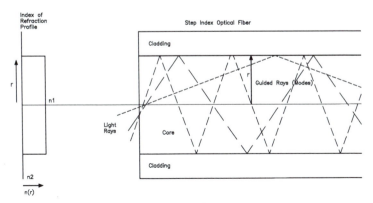

FIGURE 3.5 Light propagation in a step index fiber.

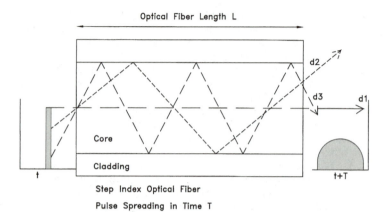

FIGURE 3.6 Step index multimode fiber dispersion.

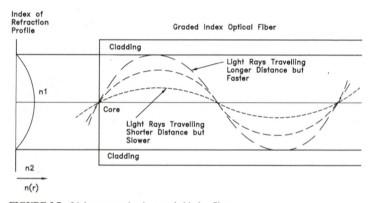

FIGURE 3.7 Light propagation in a graded index fiber.

core has a lower refractive index than the center. Light travels faster in a material with a lower index of refraction (light velocity = c/n). Light rays traveling the longer distance, in the outer core region, require more time to get to the fiber end. However, because the light travels faster in the outer region than in the core center, the longer time caused by the distance is partially compensated for by the ray's higher speed. This reduces the amount of pulse spreading between the core center and outer region light rays, thereby reducing the multimode dispersion. Thus, this type of fiber has a higher data rate bandwidth than the step index fiber.

3.4 SINGLE-MODE FIBER

A single-mode fiber is an optical fiber that only propagates one light mode (one light ray path down the center of the fiber; see Fig. 3.8). This occurs because the very small fiber core diameter, 7 to 10 μm, is near the wavelength of the light, approximately 1.26 to 1.6 μm (1260 to 1600 nm). The core acts as a wave guide for the light mode, and the cladding creates the proper boundary conditions, keeping the light mode captive and propagating in the core. This can be explained by electromagnetic wave theory and Maxwell's equations, which are beyond the scope of this book. Single-mode light wavelengths shorter than approximately 1.26 μm will not satisfy the wave equations for 7- to 10-μm fiber core size and therefore will not propagate in the fiber. The shortest wavelength of light that a single-mode fiber will propagate is called the cut-off wavelength. Below this wavelength the fiber behaves as a multimode fiber. The cut off wavelength for most fibers ranges between 1260 and 1270 nm.

The refractive index profile of single-mode fiber is similar to that of the multimode step index fiber. Because of this very small core size, it is more difficult to couple light to the fiber. A solid-state laser is commonly used to accomplish this task. Higher-precision components must also be used for all fiber connections and splices.

Because only one mode is propagated in the fiber, pulse spreading due to modal dispersion is eliminated. This enables much higher data rate transmissions over long distances. Data rates of over 10 Gbps in single-mode fiber are now common. Telephone companies are one of the primary users take advantage of this technology in order to accommodate their long-distance traffic.

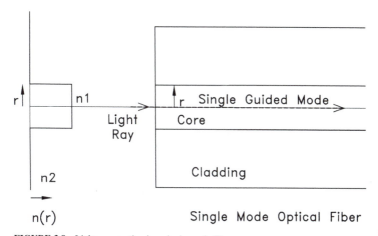

FIGURE 3.8 Light propagation in a single-mode fiber.

3.5   *OPTICAL POWER LOSS (ATTENUATION)*

Light traveling in an optical fiber loses power over distance. The loss of power depends on the wavelength of the light and on the propagating material. For silica glass, the shorter wavelengths are attenuated the most (see Fig. 3.9). The lowest loss occurs at the 1550-nm wavelength, which is commonly used for long-distance transmissions.

The loss of power in light in an optical fiber is measured in decibels (dB). Fiber optic cable specifications express cable loss as attenuation per 1-km length as dB/km. This value is multiplied by the total length of the optical fiber in kilometers to determine the fiber's total loss in dB.

Optical fiber light loss is caused by a number of factors that can be categorized into extrinsic and intrinsic losses:

- Extrinsic
  - Bending loss
  - Splice and connector loss
- Intrinsic
  - Loss inherent to fiber
  - Loss resulting from fiber fabrication
  - Fresnel reflection

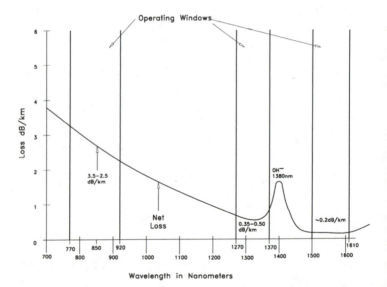

FIGURE 3.9   Optical fiber operating wavelengths.

Bend Loss. Bend loss occurs at fiber cable bends that are tighter than the cable's minimum bend radius. Bending loss can also occur on a smaller scale from such factors as:

- Sharp curves of the fiber core
- Displacements of a few millimeters or less, caused by buffer or jacket imperfections
- Poor installation practice

This light power loss, called microbending, can add up to a significant amount over a long distance.

Splice and Connector Loss. Splice loss occurs at all splice locations. Mechanical splices usually have the highest loss, commonly ranging from 0.2 to over 1.0 dB, depending on the type of splice. Fusion splices have lower losses, usually less than 0.1 dB. A loss of 0.05 dB or less is usually achieved with good equipment and an experienced splicing crew. High loss can be attributed to a number of factors, including:

- Poor cleave
- Misalignment of fiber cores
- An air gap
- Contamination
- Index-of-refraction mismatch
- Core diameter mismatch

to name just a few.

Losses at fiber optic connectors commonly range from 0.25 to over 1.5 dB and depend greatly on the type of connector used. Other factors that contribute to the connection loss include:

- Dirt or contaminants on the connector (very common)
- Improper connector installation
- A damaged connector face
- Poor scribe (cleave)
- Mismatched fiber cores
- Misaligned fiber cores
- Index-of-refraction mismatch

Loss Inherent to Fiber. Light loss in a fiber that cannot be eliminated during the fabrication process is due to impurities in the glass and the absorption of light at the molecular level. Loss of light due to variations in optical density, composition, and molecular structure is called Rayleigh scattering. Rays of light encountering these variations and impurities are scattered in many directions and lost.

The absorption of light at the molecular level in a fiber is mainly due to contaminants in glass such as water molecules (OH^-). The ingress of OH^- molecules into an optical fiber is one of the main factors contributing to the fiber's increased attenuation in aging. Silica glass's (SiO_2) molecular resonance absorption also contributes to some light loss.

Figure 3.9 shows the net attenuation of a silica glass fiber and the three fiber operating windows at 850, 1310, and 1550 nm. For long-distance transmissions, 1310- or 1550-nm windows are used. The 1550-nm window has slightly less attenuation than 1310 nm. The 850-nm communication is common in shorter-distance, lower-cost installations.

Loss Resulting from Fiber Fabrication.   Irregularities during the manufacturing process can result in the loss of light rays. For example, a 0.1 percent change in the core diameter can result in a 10-dB loss per kilometer. Precision tolerance must be maintained throughout the manufacturing of the fiber to minimize losses.

Fresnel Reflection.   Fresnel reflection occurs at any medium boundary where the refractive index changes, causing a portion of the incident light ray to be reflected back into the first medium. The fiber end is a good example of this occurrence. Light, traveling from air to the fiber core, is refracted into the core. However, some of the light, about 4 percent, is reflected back into the air. The amount being reflected can be estimated using the following formula:

Reflected Light Power at a Boundary

Reflected light (%) = $100 \times (n1 - n2)^2/(n1 + n2)^2$

where $n1$ = the core refractive index
$n2$ = the air refractive index

At a fiber connector, the light reflected back can easily be seen with an optical time domain reflectometer (OTDR) trace (see Fig. 13.2). It appears as a large upward spike in the trace. This reflected light can cause problems if a laser is used and should be kept to a minimum (see Sec. 12.5).

The reflected light power can be reduced by using better connectors. Connectors with the "PC" (Physical Contact) or "APC" (Angle Physical Contact) designations are designed to minimize this reflection.

3.6 FIBER BANDWIDTH

Optical fiber bandwidth is a measure of the information-carrying capacity of an optical fiber. The optical fiber bandwidth is limited by the fiber's total dispersion (pulse-broadening) and nonlinear effects. Dispersion limits information-carrying capacity because pulses distort and broaden, overlapping one another and becoming indistinguishable to the receiving equipment. Nonlinear components also cause pulse distortion and wavelength interference. To reduce the distortion, pulses must be transmitted less frequently (thereby reducing the data rate).

As Fig. 3.10a shows, the original optical data pulses are discrete—ones and zeros that can be easily identified. After the signal propagates in the optical fiber for some distance (Fig. 3.10b), dispersion occurs. The pulses widen but can still be decoded by receiving equipment.

Additional dispersion can introduce errors into the transmission. After further propagation in the optical fiber (Fig. 3.10c), the signal becomes totally distorted, and receiving equipment cannot derive the original waveform. Data transmission is not possible.

a. Digital optical signal with no dispersion. Ideal signal.

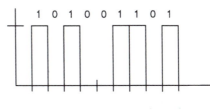

b. Some dispersion to the same signal, but still acceptable.

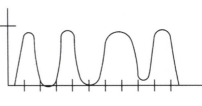

c. Extensive dispersion to optical signal. Individual pulses are not distinguishable. This signal is not acceptable.

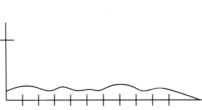

FIGURE 3.10 Signal dispersion.

Also, as dispersion increases, the optical signal peak power is reduced, which affects the receiver's optical budget (see Sec. 12.6). Dispersion is a function of optical fiber length; the longer the fiber, the more pronounced the effect. As long as the optical fiber dispersion is within the lightwave equipment's specifications, fiber optic transmission is possible.

Dispersion also affects analog optical signals by distorting the signal. It should be considered for all analog designs such as for TV channel transmission for the cable TV industry.

Total dispersion can be divided into three categories: chromatic dispersion, polarization mode dispersion, and multimode dispersion (also called modal dispersion). Chromatic dispersion can be further subdivided into chromatic waveguide dispersion and chromatic material dispersion, as graphically represented here:

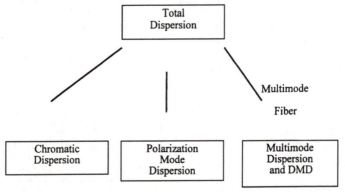

Multimode Dispersion.   Multimode dispersion, also known as modal dispersion, is measured in nanoseconds per kilometer, affects only multimode fiber and is caused by different paths, or modes, that a light ray follows in the fiber (see Fig. 3.6). This results in light rays traveling different distances and arriving at the other end of the fiber at different times. A transmitted pulse will broaden because of this effect, and this consequently reduces the maximum effective data rate.

Total pulse spreading Δtm is:

Dtm 5 model dispersion (ns/km) 3 fiber length (km)

Step index fiber has the highest modal dispersion and consequently the lowest bandwidth. Because of the nonuniform profile of index of refraction in a graded index fiber, the modal dispersion is decreased. This results in a transmission rate, which is higher than that in a step index fiber.

Differential Mode Delay (DMD).   Differential mode delay (DMD) is delay between paraxial and skew rays in a multimode fiber. It occurs when a laser source (designed to be used with single-mode fiber) is used to launch light into a multimode fiber. The result is DMD-induced jitter and a substantial decrease in the multimode fiber bandwidth (40 to 50 percent can occur). However, this jitter can be significantly reduced by laterally offsetting the laser launch into the multimode fiber. Offset launch patch cords or adapters are available for this purpose. These offset launch patch cords (also called mode conditioning patch cords) should be used whenever a single-mode fiber laser is used with a multimode fiber. Recently released, specialized multimode fiber types, which do not require these offset patch cords when using laser sources, can be deployed. The fiber manufacturer should be consulted for more details.

Chromatic Dispersion. The speed of light in a fiber is dependent on the fiber's refractive index n as shown by formula $v = c/n$. However, n changes with wavelength. Two light rays of different wavelengths in the same fiber will have different refractive indexes and consequently have different speeds in the fiber. A fiber light source is composed of a spectrum of many wavelengths, and hence the source spectrum constituents will be propagated through the fiber at different speeds. Each will arrive at its destination at slightly different times, thereby resulting in pulse spreading. A typical plot of a fiber's chromatic dispersion characteristic is shown in Fig 3.11.

Chromatic dispersion is measured in units picoseconds per kilometer (of fiber) per nanometer (of light source spectral width). Pulse spreading due to chromatic dispersion is:

Δtc = dispersion (ps/nm.km) $\times$ fiber length (km) $\times$ spectral width (nm).

As shown in Fig 3.11, a light source of wavelength λ has dispersion values D. A light pulse from this source will disperse in time during propagation in the fiber. This dispersion can be compensated, either by adding proper dispersion compensating fibers or modules, or by reducing the laser spectral width.

Polarization Mode Dispersion (PMD). This single-mode fiber term represents the dispersion of optical light caused by two orthogonal states of polarized

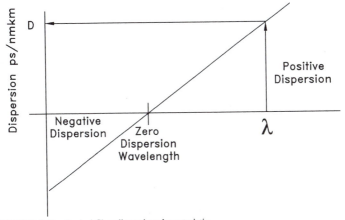

FIGURE 3.11 Typical fiber dispersion characteristic.

light propagating in a fiber. Both of these states propagate at slightly different velocities and thereby arrive at their destination at different times. PMD is generally not an issue for systems communicating at a data rate less than 2.5 Gbps (OC-48 / STM-16) or short distances. However it can cause significant problems for OC-192 systems transmitting over long unrepeated (using optical amplifiers) fiber distances. PMD should be measured for all long-haul OC-48 or OC-192 fiber installations.

PMD is affected by the geometry of a fiber and random changes to a fiber due to fiber bends, temperature changes, and other stresses on the fiber. Because the effects can be random, successive measurements of PMD on the same fiber can vary.

Fiber manufacturers will specify PMD with units of ps/$\sqrt{km}$ (picoseconds/$\sqrt{kilometer}$) at the time of manufacture. This value can, however, change after it has been installed. Due to improved fiber manufacturing techniques, new fiber types have generally lower PMD values than older fiber types.

Pulse spreading due to PMD is:

$$\Delta tp = PMD \ (ps/\sqrt{km} \times \sqrt{fiber \ length \ (km)}$$

PMD accumulates over the fiber length and the overall PMD delay is referred to as differential group delay (DGD), which is field-measured in units of picoseconds (ps). DGD can be estimated by the formula below using manufacturers' PMD values, but it is better to field-measure it, using available PMD testers, because of its random nature.

$$Mean \ link \ DGD = [\Sigma(PMD)^2 \times span \ length]^{1/2}$$

DGD   ps

PMD   ps/$\sqrt{km}$

Span Length   km

DGD Calculation Example:

Four sections of fiber cable are spliced together to create an OC-192 system link. The cables lengths and PMD values provided by the manufacturer are as follows:

Section A — 0.2 ps/$\sqrt{km}$, length 20 km

Section B — 0.15 ps/$\sqrt{km}$, length 10 km

Section C — 0.22 ps/$\sqrt{km}$, length 30 km

Section D — 0.11 ps/$\sqrt{km}$, length 7 km

Mean link DGD = $[0.2^2 \times 20 + 0.15^2 \times 10 + 0.22^2 \times 30 + 0.11^2 \times 7]^{1/2}$

Mean link DGD = 1.6 ps

Field-measuring fiber link DGD should be done a number of times over a period using worst-case results. Generally, DGD is not an issue unless systems with data rates of 10 Gbps (OC-192/STM-64) or higher are used. Once the link DGD is known, it is compared to manufacturer's specified DGD to obtain a link budget derating in decibel called polarization dependent loss (PDL). PDL is the peak-to-peak optical power variation when the fiber is exposed to all states of polarization. In some cases, improvement of the link PDL may be required for the link to operate.

Nonlinear Effects.   Nonlinear effects such as four-wave mixing, stimulated Brillouin scattering (SBS), self-phase modulation, and cross-phase modulation can also impair system operation. These effects can cause problems when long, unregenerated fiber distances and high optical power levels are used.

Four-wave mixing can occur in DWDM networks. It can cause various wavelengths in a fiber to interact and interfere with each other. This effect is worst in fibers with very low chromatic dispersion such as dispersion shifted fiber (DSF). Constant chromatic dispersion along the fiber's length (either positive or negative) helps to suppress these nonlinear components.

A new fiber type called nonzero dispersion shifted fiber (NZDSF) was developed to reduce these effects. NZDSF fiber maintains a small amount of chromatic dispersion along its length, which helps to suppress four-wave mixing components. The use of uneven channel spacing or increased space between channels can also help reduce four-wave mixing.

Total Bandwidth and Dispersion.   Multimode fiber bandwidth is specified by the fiber manufacturer in the normalized modal bandwidth distance product form "megahertz $\times$ kilometer" (MHz $\times$ km). This bandwidth product accounts only for pulse spreading due to multimode (modal) dispersion. To determine the optical fiber's total bandwidth, chromatic dispersion effects should also be considered.

Total bandwidth of multimode fiber can be calculated as follows:

$$B_{Total} = (B^{-2}_{Modal} + B^{-2}_{Chromatic})^{-1/2}$$

Total dispersion of multimode fiber can be calculated by:

$$\Delta t_{mm} = \sqrt{(\Delta tm)^2 + \Delta t_c)^2}$$

However, in many multimode applications, modal dispersion is the predominant dispersion effect.

Single-mode fiber bandwidth is limited by the fiber's chromatic dispersion [which is specified in the form of picoseconds/nanometers $\times$ kilometers) or (ps/nm.km)], PMD, and nonlinear effects. Chromatic dispersion changes with wavelength and can be positive, negative, or zero.

Total chromatic and PMD dispersion of single mode fiber can be calculated by:

$$\Delta t_{sm} = \sqrt{(\Delta tp)^2 + (\Delta tc)^2}$$

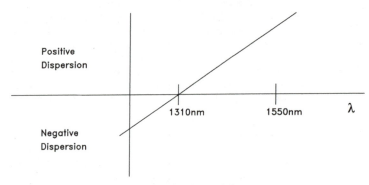

FIGURE 3.12 NDSF chromatic dispersion characteristic.

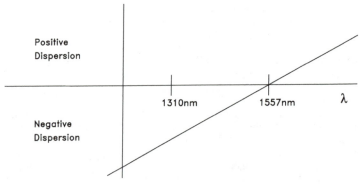

FIGURE 3.13 DSF chromatic dispersion characteristic.

A fiber is designed with specific characteristics that set its chromatic dispersion for wavelengths. The most common type of fiber is called a nondispersion shifted fiber (NDSF), which has a dispersion characteristic similar to Fig 3.12 [it is also called standard single-mode fiber (SSMF)]. This fiber has zero dispersion at 1310 nm and positive dispersion at 1550 nm.

A second, once popular, type of fiber is called dispersion shifted fiber (DSF). This fiber is manufactured with characteristics that move its zero dispersion wavelength to the 1550-nm band (around 1557 nm) instead of 1310 nm (see Fig 3.13).

Having zero chromatic dispersion in the 1550-nm window is desirable because these wavelengths are used for long-distance communications and low dispersion will help increase transmission distance. Unfortunately, if wavelength division multiplexing (WDM) is deployed in DSF fiber, then nonlinear components can occur in the fiber which can interfere with the transmission (four-wave mixing).

A new fiber type which has been developed is called near-zero dispersion shifted fiber (NZ-DSF) and helps to suppress these unwanted components. The NZ-DSF fiber maintains a small amount of chromatic dispersion along its length that is used to suppress the effect of four-wave mixing components. NZ-DSF fiber, because of its low dispersion at 1550 nm, provides better long-distance transmission characteristics than regular NDSF fiber and helps to suppress four-wave mixing components.

Excessive chromatic dispersion can also be compensated at the fiber ends. If a fiber is found to have excessive positive dispersion, dispersion compensating fiber with negative dispersion can be added to counter the positive dispersion. Dispersion compensating modules (DCMs) are also available. NDSF fiber will always have positive dispersion in the 1550-nm band, so negative compensating dispersion may need to be added to the link. DSF fiber will have positive, negative, or zero dispersion, depending on the wavelength, and positive or negative dispersion compensation may be required.

In selected lightwave equipment, optical transmitter lasers can be adjusted so that the laser light will also counteract excessive positive or negative fiber dispersion. This is called adding laser chirp (positive or negative). Since this compensation may not be sufficient for some fiber links, DCMs are used.

3.7 SOLITON TRANSMISSION

A soliton is a type of optical wave, composed of narrow pulses of light, that retains its shape as it travels over long distances in the fiber. Because of its

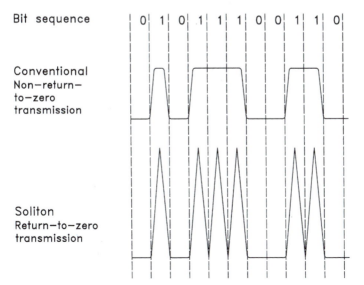

FIGURE 3.14 Soliton optical signal.

shape, chromatic dispersion and nonlinear distortion have less effect on this wave than on standard SONET waves.

Soliton transmission uses a return-to-zero encoding scheme as opposed to the nonreturn-to-zero encoding used for SONET transmissions (see Fig. 3.14). Soliton transmission has, in the lab, been tested error-free for distances of 2000 km using only optical amplifiers to boost the signal. Results have also shown that signal regeneration is not required. This type of optical transmission allows for much greater distances between optical regenerators and higher data rates over all types of fiber. Soliton transmission equipment is not yet commercially available, but companies are testing products in the lab for possible future deployment.

3.8 OPTICAL FIBER SPECIFICATION: AN EXAMPLE

The following is an example of a typical optical fiber specification sheet available from a fiber manufacturer:

Specification	Unit	Explanation
Core diameter	μm	Fiber core diameter
Cladding diameter	μm	Cladding diameter does not include coating
Coating diameter	μm	Includes plastic colored coating
Mode field diameter	μm	This value is used for single-mode fibers only.
Attenuation at: 850 nm	dB/km	Attenuation per kilometer (for multi-mode fiber only)
Attenuation at: 1310 nm	dB/km	Wavelength available for both multi-mode and single-mode fiber
Attenuation at: 1550 nm	dB/km	Lower attenuation at longer wavelength (single-mode fiber only)
Fiber optic bandwith: 850 nm	MHz × km	Multimode fiber's modal bandwidth at wavelength
Fiber optic bandwidth: 1310 nm	MHz × km	Multimode fiber's modal bandwidth at wavelength
Chromatic dispersion	ps/nm × km	Should be specified for operating wavelength
PMD	ps//√km	Polarization mode dispersion
Cut-off wavelength	1260–1270 nm	Shortest wavelength the fiber will propagate (single-mode fibers only)
Fiber manufacturer	XY company	Name of optical fiber manufacturer (not necessarily the same as fiber cable manufacturer)

CHAPTER 4
FIBER OPTIC CABLES

Fiber optic cables are available for either outdoor or indoor environments. The outdoor cable must be capable of withstanding climate extremes. It must operate in a wide temperature range, block water ingress, withstand the sun's ultraviolet radiation, not fail in heavy winds or other mechanical stresses, and resist gnawing rodents. It is built to be rugged with heavy jacket and often metal armor. It should also have a high packing density to maximize the number of fibers in the cable.

The indoor cable is often placed in a more controlled environment. Although it does not need to be as rugged as the outside plant cable, it still needs to protect the fibers from inside elements such as mechanical stress during and after installation. It also needs to be more flexible to allow for installation throughout building raceways and easier termination with connectors. It also must meet or exceed the National Electrical Code (NEC) and any other building codes.

4.1   OUTDOOR CABLES

4.1.1   Loose Tube Cable

Loose tube fiber optic cable is designed primarily for outdoor installation. It is comprised of a number of fiber tubes surrounding a central strength member, with an overall protective jacket (see Fig. 4.1).

The distinguishing features of this type of cable are the fiber tubes. Each tube, 2 to 3 mm in diameter, carries a number of optical fibers that lie loose in the tube. The tubes can be hollow or, more commonly, filled with a special water-resistant gel that helps keep moisture away from the fiber. The loose tube also isolates the fiber from the exterior mechanical forces placed on the cable (see Fig 4.2).

The fibers in the tube are slightly longer than the cable itself so that the cable can elongate under tensile loads without applying stress to the fiber.

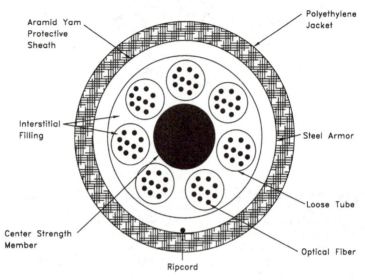

FIGURE 4.1 Loose tube cable.

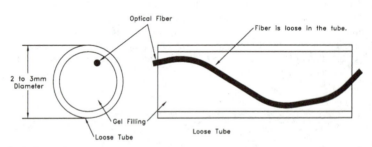

FIGURE 4.2 Fiber optic cable loose tube.

During OTDR testing, this excess fiber length should be factored in to determine a more precise physical length for the cable (the excess fiber length is available from the manufacturer).

Each tube is colored, or numbered, and each individual fiber in the tube is, in turn, colored for easier identification. The number of fibers available in each cable range from a few to over 200.

The center of the cable contains the strength member, which can be steel, Kevlar, or a similar material. This member provides the cable with strength and support during pulling operations as well as in permanently installed positions. It should always be securely fastened to the pulling eye during cable-pulling operations, and to appropriate cable anchors in splice enclosures or patch panels.

The cable jacket can be made from polyethylene, steel armor, rubber, and aramid yarn, among other materials, for various environments. For easier and more accurate OTDR fault locating, the jacket is sequentially numbered at every meter (or every foot) by the manufacturer. The pulling tension and bending radius of fiber optic cables vary, so the manufacturer's specifications should be consulted for individual cable details.

Loose tube cables are used for most outdoor installations, including aerial, duct, and direct buried applications. Loose tube cable is not suitable for installation in high vertical runs because of the possibility of gel flow or fiber movement. Manufacturer's specifications should be consulted to determine the cable's maximum vertical rise for all installations.

4.1.2   Figure 8 Cable

The Figure 8 cable is a loose tube cable with a factory-attached messenger. A messenger is the support member used in aerial installations. It is usually a high-tension steel cable with a diameter between $1/4$ and $5/8$ in. The Figure 8 cable is aptly named because its cross-sectional design resembles the number 8 (see Fig. 4.3). It is used for aerial installations and eliminates the need to lash the cable to a preinstalled messenger. Aerial installation of fiber optic cable is much quicker and easier with the Figure 8 cable.

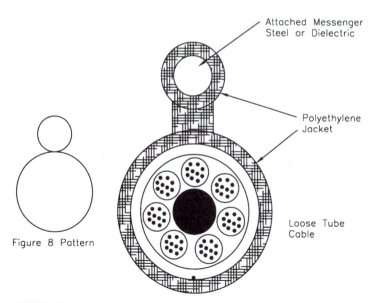

FIGURE 4.3   Figure 8 fiber optic cable.

The messenger is available in a high-tension steel or an all-dielectric material. The dielectric messenger should be considered when installing the cable near high-voltage power lines.

4.1.3 Armored Cable

Armored cables have a protective steel armor beneath a layer of polyethylene jacket (see Fig. 4.4). This provides the cable with excellent crush-resistance and rodent-protection properties. It is commonly used in direct burial applications or for installation in heavy industrial environments. The cable is usually loose tube, but a tight-buffered type is also available. The steel armor also provides added protection against water ingress into the cable fibers. This increases the useful life of the cable fibers.

Armored cable is also available with a double-armor protective jacket for added protection in harsh environments. The steel armor should always be properly grounded to an earth ground at all termination points, splice locations, and all building entrances.

4.1.4 Ribbon Cable

Fiber optic ribbon cable was introduced into the market as a solution to increase fiber densities in cables and decrease the time required to prepare

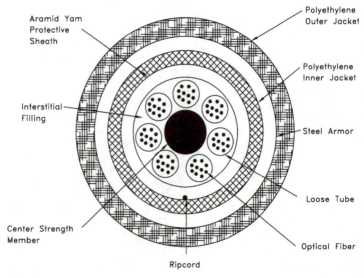

Aramid Yarn Protective Sheath

Interstitial Filling

Center Strength Member

Ripcord

Polyethylene Outer Jacket

Polyethylene Inner Jacket

Steel Armor

Loose Tube

Optical Fiber

FIGURE 4.4 Armored fiber optic cable.

and splice high-fiber-count cables. The solution was to bundle a group of fibers in a row that resembled a ribbon (hence the name) (see Fig 4.5). The ribbon contains 12 or 24 fibers that are placed in a cable's buffer tubes. This allows for an increase in cable densities from the traditional 200 to 300 fiber cables to 800 or more fibers in a ribbon cable.

There are two basic cable constructions for ribbon cable: single central tube and stranded loose tube cable.

The single central tube construction contains one loose tube in the center of the cable that houses the fiber ribbons. The 12 fiber ribbons are stacked in the central tube and surrounded by a filling compound. As in a regular loose tube cable, the individual fiber ribbons are loose in the tube, thereby isolating them from stress on the cable jacket. Cable strength members are dielectric or metallic rods placed on each side of the central tube. The cable coating is either a polyethylene jacket or a metal armor with a polyethylene cover (see Fig 4.6).

The stranded loose tube cable is similar to the construction of the traditional loose tube cable. However, the individual loose fibers in each buffer tube are replaced by one or more 12-fiber ribbons (see Fig 4.7).

The following identifies some advantages and disadvantages of deploying ribbon cable.

Advantages

1. High-fiber-density ribbon cables save space in ducts.

2. Using a mass fusion splicer, and proper ribbon cable stripping and cleaving tools, splicing time per fiber is less than with standard cable of equal fiber count.

3. Longer cable lengths can be placed on a cable reel, due to the higher density of the cable.

Disadvantages

1. A special mass fusion splicer and other fiber ribbon stripping and cleaving tools are required. Special training is required for splicing crews.

2. Special splice enclosures are required to accommodate the ribbon cable.

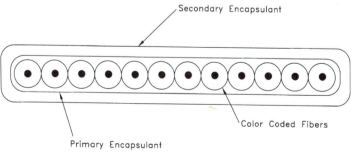

FIGURE 4.5 Fiber ribbon.

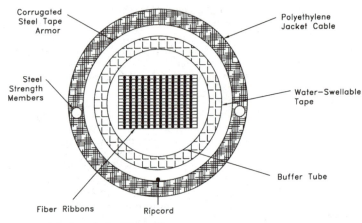

FIGURE 4.6 Central tube ribbon cable.

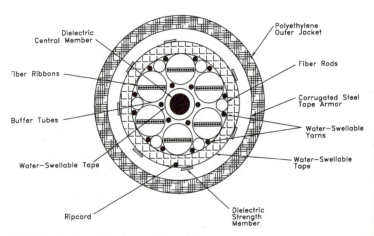

FIGURE 4.7 Stranded loose tube ribbon cable.

4.2 INDOOR CABLES

4.2.1 Tight-Buffered Cable

Tight-buffered fiber optic cable is designed for indoor environments. It contains a number of individually buffered fibers surrounding a central strength member, with an overall protective jacket (see Fig. 4.8).

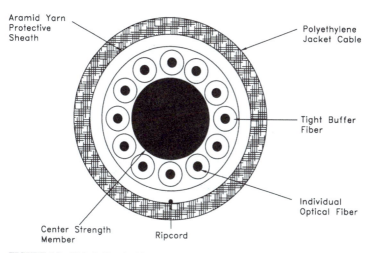

FIGURE 4.8 Tight-buffered cable.

The fiber buffer is a 900-μm-diameter plastic cover surrounding the optical fiber's 250-μm coating (see Fig. 4.9).

The buffer provides each individual fiber with protection from the environment as well as physical support. Each individual buffered fiber can be run outside of the cable jacket for short distances and connectorized directly (connector installed directly onto cable fiber), without the need for a splice tray. For some installations, this can decrease installation costs and reduce the number of splices in a fiber length. Because of the tight-buffer design of the cable, it is more sensitive to tensile loading and bends, and can have increased microbending loss.

Tight-buffered cable is more flexible than most loose tube cable and thus has a smaller bending radius. It can be installed in higher vertical rises than can loose tube cable because of the individual fiber buffer support. However, because of this support, it is comparatively larger in diameter and usually more expensive than a similar fiber count loose tube cable.

4.2.2 Fan-Out Cable

A fan-out cable is a tight-buffered cable terminated with connectors. Each individual fiber has either a 900-μm or 3-mm jacket, depending on the application. The 900-μm jacket is deployed where the cable will be terminating in a fiber patch panel, and each fiber will remain protected by the patch panel enclosure. The 3-mm type provides additional protection for the fibers and is used where the cable connects directly to the lightwave equipment.

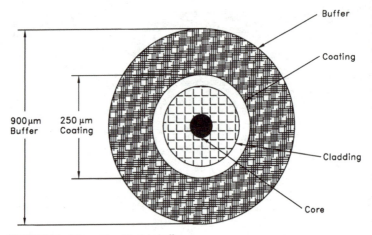

FIGURE 4.9   Fiber optic cable tight buffer.

Because of the thicker jacket on each fiber, the 3-mm-type cable has a smaller fiber count for same-size cable diameter. The cable is usually purchased with connectors mounted in the factory, which results in a higher-quality connector installation and lower connector loss.

The 900-μm-type fan-out cable is best used for fiber connectivity between splice enclosure and fiber patch panel (See Fig 4.10).

4.2.3   Fiber Optic Patch Cords (Jumpers)

Fiber optic patch cords, also known as fiber jumpers, are short lengths of

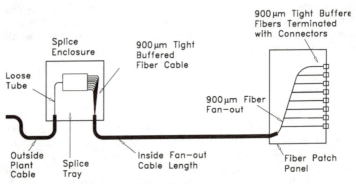

FIGURE 4.10   Typical fan-out cable application.

single- or dual-fiber cable. Each fiber is protected with a 3-mm jacket and has connectors at both ends. A yellow-jacketed jumper indicates a single-mode fiber, while an orange-jacketed jumper indicates that it is a multimode fiber. These patch cords are used to connect lightwave equipment to fiber patch panels, or connect fiber equipment together. They can also be cut in half and used as pigtails to terminate other fiber cables. Fiber patch cords have much tighter bending radii then any other cable, which is suitable for their application. Their bend radius for unloaded installation should be kept larger than 3 cm (1.2 in) or larger than 5 cm (2 in) for loaded installation.

4.3 OTHER CABLES

Other fiber optic cables are available for special applications.

4.3.1 Self-Supporting Aerial Cable

Self-supporting aerial cable is a loose tube cable designed for use in aerial installations. It does not require a messenger for support. The cable is designed with a heavy strength member and jacket, usually Kevlar that can withstand its own load and other stresses during and after installation. Special clamps are used to secure the cable directly to pole structures. A special version of this cable, called the all-dielectric self-supporting cable (ADSS), is often used in electric utility installations. Being completely dielectric, the cable can be installed directly underneath high-voltage lines.

4.3.2 Submarine Cable

Submarine cable is a loose tube cable designed for underwater submersion. It has a heavy armor jacket and may contain copper pairs to provide power for submersed optical amplifiers. Many continents are now connected by transoceanic fiber optic submarine cable.

4.3.3 Optical Ground Wire (OPGW)

Optical ground wire (OPGW) cable is an electric utility ground wire cable (found at the very top of high voltage towers) that has optical fibers inserted into the center core of the cable. The optical fibers are completely protected by metallic tube that is surrounded by heavy-gauge ground wire. It is used by electric utilities to provide communications along high-voltage cable routes.

4.4   CABLE FIRE RATING

The National Electrical Code (NEC) requires that all fiber optic cable used indoors be marked correctly and installed properly for its intended use. NEC specifies that all indoor cables be marked with appropriate fire and smoke ratings. Three different building regions are identified by the NEC: plenums, risers, and general-purpose areas.

- A plenum is a building space, compartment, duct, or chamber used for air flow or to form part of an air distribution system.
- A riser is a floor opening, shaft, or duct that runs vertically through one or more floors.
- A general-purpose area is all other area that is not plenum or riser, which is on the same floor.

   A summary of NEC cable markings is listed below. The NEC should be consulted before proceeding with any indoor cabling installation.

Cable marking	Name	UL/CSA test	Possible substitute
OFNP	Optical Fiber Nonconductive Plenum Cable	UL-910	
OFCP	Optical Fiber Conductive Plenum Cable	UL-910	
OFNR	Optical Fiber Nonconductive Riser Cable	UL-1666	OFNP
OFCR	Optical Fiber Conductive Riser Cable	UL-1666	OFCP
OFNG	Optical Fiber Nonconductive General-Purpose Cable	CSA C22.2	OFNP OFNR
OFCG	Optical Fiber Conductive General-Purpose Cable	CSA C22.2	OFCP OFCR
OFN	Optical Fiber NonConductive	UL-1581	OFNP OFNR
OFC	Optical Fiber Conductive	UL-1581	OFCP OFCR

OFNP Cable.   Optical Fiber Nonconductive Plenum Cable can be installed in ducts, plenums, and other spaces used for building airflow. This cable has fire-resistance and low smoke production characteristics. This is the highest cable fire rating, and no other cable types can be used as substitutes. This cable type can be marked as "FT-6" by Canadian Standards Association (CSA).

OFCP Cable.   Optical Fiber Conductive Plenum Cable has the same fire rating characteristics as OFNP cable. It has a conducting armor and/or central strength member, usually steel. It should be properly grounded at both ends. The steel armor provides excellent protection for the cable in harsh environments or industrial installations. As a conductor, it cannot be placed in the same cable tray or conduit as power cables.

OFNR Cable. Optical Fiber Nonconductive Riser Cable can be installed in building vertical shafts (risers) or in runs from one floor to another floor. It cannot be installed in plenums. This cable has fire-resistance characteristics tested to UL-1666 "Standard Test for Flame Propagation Height of Electrical and Optical Fiber Cable Installed Vertically in Shafts." OFNP-type cable can be used as a substitute for this cable.

OFCR Cable. Optical Fiber Conductive Riser Cable has the same fire rating characteristics as OFNR cable. It has a conducting armor and/or central strength member, usually steel. It should be properly grounded at both ends. The steel armor provides excellent protection for the cable in harsh environments or industrial installations. As a conductor, it cannot be placed in the same cable tray or conduit as power cables.

OFNG Cable. Optical Fiber Nonconductive General-Purpose Cable can be installed in typical horizontal, single-floor installations. It cannot be installed in plenums or risers. This cable type is equivalent to the Canadian CSA "FT-4" rating. OFNP or OFNR type cables can be used as a substitute for this cable.

OFCG Cable. Optical Fiber Conductive General-Purpose Cable has the same fire rating characteristics as OFNG. It has a conducting armor and/or central strength member, usually steel. It should be properly grounded at both ends. The steel armor provides excellent protection for the cable in harsh environments or industrial installations. As a conductor, it cannot be placed in the same cable tray or conduit as power cables.

OFN Cable. Optical Fiber Nonconductive Cable is also a general-purpose cable that can be installed in typical horizontal, single-floor installations. It cannot be installed in plenums or risers. This cable type does not have the equivalent CSA rating. OFNP or OFNR type cables can be used as a substitute for this cable.

OFC Cable. Optical Fiber Conductive Cable has the same fire rating characteristics as OFN. It has a conducting armor and/or central strength member, usually steel. It should be properly grounded at both ends. The steel armor provides excellent protection for the cable in harsh environments or industrial installations. As a conductor, it cannot be placed in the same cable tray or conduit as power cables.

Outside Plant and Unmarked Cables. NEC allows outside plant cables or other unmarked cables to be installed indoors as long as they are completely installed in metal conduit or electrical metallic tubing.

NEC also allows unmarked outside plant cables to enter a building for up to 50 ft beyond the entrance point before the cable needs to be terminated (see Figure 4.11). This relief provides for reasonable cable length at a cable entrance to convert from a non-fire-resistant outside plant cable to a properly marked fire-resistant cable.

For Canadian installations, all indoor cables must be marked with the proper CSA FT rating before they can be installed indoors. NEC cable marking is not sufficient.

Before installing any indoor cable, the NEC should be consulted. Local building codes should also be checked because cables that meet NEC codes may not always meet local building code requirements.

The table given previously provides a general guide for fiber optic cable applications.

The fiber optic cable manufacturers for each of these cable types should be contacted for further product details.

4.5 CABLE COMPOSITION

Fiber optic cables are constructed with various materials to suit the installation environment. Careful consideration of cable composition is essential to prolonging cable life.

Outdoor cables should be strong, weatherproof, and ultraviolet (UV)-resistant. The cable should be able to withstand the maximum temperature

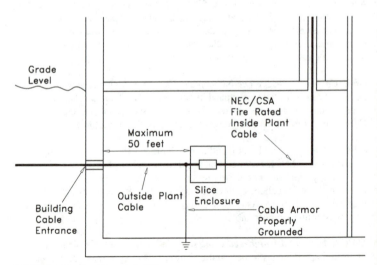

FIGURE 4.11 Outside cable building entrance.

variations that can be encountered during the installation process and throughout its life span. Often a cable is specified with two temperature ranges. One range specifies acceptable cable-handling and installation temperatures, and the other indicates the cable's maximum temperature range after it has been installed and is in its final static position.

Outdoor cables are treated to inhibit UV light from penetrating the jacket and causing decomposition of the internal material. Jackets can be specified with additional UV protection if required.

Indoor cables should be strong and flexible with the required flame-retardant and smoke-inhibitor rating. Jacket colors can be bright orange or yellow for easy identification. Some of the more popular cable materials are listed in the following sections.

Polyethylene (PE). Polyethylene is a popular cable jacket for outdoor installations. The black-jacket type has good moisture- and weather-resistance properties. It is a very good insulator and has stable dielectric properties. Depending on its molecular density, it can be very hard and stiff, especially in colder temperatures. Alone it is not flame retardant, but it can be if treated with the proper chemicals.

Polyvinyl Chloride (PVC). PVC jackets offer good resistance to environmental effects, with some formulations rated for temperatures of $-55°$ to $+55°C$. It has flame-retardant properties and can be found in outdoor as well as indoor installations. PVC is less flexible than PE and usually more expensive.

Polyfluorinated Hydrocarbons (Fluoropolymers). Certain formulations of polyfluorinated hydrocarbon jacket material have good flame-resistance properties, low smoke characteristics, and good flexibility. It is used for indoor installations.

Aramid Yarn/Kevlar. Aramid yarn is a light material found just inside the cable jacket surrounding the fibers and can be used as a central strength member. The material is strong and is used to bundle and protect the individual tubes or fibers in the cable. Kevlar, a particular brand of aramid yarn, has very high strength and is often used in bulletproof vests. Fiber optic cables that must withstand high pulling tensions often use Kevlar as a central strength member. When placed just inside the cable jacket, where it surrounds the entire cable interior, Kevlar provides the fibers with additional protection from the environment. It can also provide bullet-resistance properties to the cable, which may be required in aerial cable installations running through hunting areas.

Steel Armor. A steel armor jacket is often used for outdoor and indoor installations. When used in a direct burial cable, it provides excellent crush resistance and is the only material that is truly rodent-proof. For industrial environments it is used inside the plant when the cable is installed without conduits or cable tray protection. However, the steel added to the cable makes it a conductor, thereby sacrificing the dielectric advantage of the cable. Steel armor cables should always be properly grounded. The steel armor also decreases water ingress into the fiber, thereby increasing the life expectancy of the cable.

Ripcord.   The cable ripcord is a strong, thin thread found just below the cable jacket. It is used to easily split the cable jacket without harming the cable interior.

Central Member.   The central member is used to provide strength and support to the cable. During cable-pulling operations it should be secured to the pulling eye. For permanent installations, it should be attached to the central member anchor of the splice enclosure or patch panel.

Interstitial Filling.   This is a gel-like substance found in loose tube cables. It fills buffer tubes and cable interiors, making the cable impervious to water. It should be completely cleaned off with special gel-removing compound when the cable end is stripped for splicing.

4.6   CABLE CRUSH PERFORMANCE STANDARD

The crush resistance of a cable specifies the ability of the cable to withstand compression force as can be found in direct burial applications. The industry standard procedure for testing a cable's crush resistance can be found in document EIA-455-41 "Compressive Loading Resistance of Fiber Optic Cables" (FOTP-41). This is only a test procedure and does not specify the pass/fail criteria.

The Insulated Cable Engineers Association Inc. (ICEA) has two documents that specify criteria for inside and outside fiber optic cables, ICEA S-83-596 "Standard for Premises Distribution Cable" and ICEA S-87-640 "Standard for Fiber Optic Outside Plant Communications Cable." Both ICEA standards specify the compressive load, the amount of time the cable is under load, and fiber attenuation change while under load and after the load is removed.

4.7   FIBER OPTIC CABLE SPECIFICATION: AN EXAMPLE

The following is an example of a fiber optic cable specification sheet available from a cable manufacturer:

Specification		Explanation
Cable type	Loose tube	
Number of fibers	18	3 active tubes, 6 fibers per tube
Nominal weight:	166 kg/km	166 kg per km of cable
Diameter:	14.4 mm	Usually varies by 5 percent
Temperature range:		
Storage	−40 to 70°C	Storing cable on reel
Operating	−40 to 70°C	Installed operating
Installation	−30 to 50°C	During installation and handling
Maximum tensile rating:		
Installation	2700 N	Maximum during installation
Permanent	600 N	Operational, no measurable change in attenuation
Minimum bend radii:		
Installation	22.5 cm	While cable is being installed
Permanent	15.0 cm	Operational, no measurable change in attenuation
Crush resistance:	220 N/cm	Operational, no measurable change in attenuation
Maximum rise:	247 m	Requires no intermediate tie-downs or loops
Copper pairs:	None	Used for communications during installation or repair
Jacket:	Steel Armor	
Fire rating:		NEC rating for indoor cables
Central member:	Dielectric	Central strength member
Application	OSP direct burial	Outside plant direct burial cable

Application	Patch cords	Fan-out cable	Tight-buffered dielectric	Loose-tube dielectric	Loose tube armor	Tight-buffered armor	Figure 8	Self-support	Submarine
Direct connection to equipment in same room or cabinet[1,2]	X		X						
Terminated by patch panel[2]	X	X	X						
Terminated by splice enclosure		X	X	X	X	X	X	X	X
Between offices, intrabuilding[2]	X	X	X						
Between Buildings, interbuilding				X	X		X	X	
Inside an industrial plant[2]		X	X	X	X	X	X		
High vertical rises		X	X						
Aerial interbuilding				X			X	X	
Underground in ducts				X	X				
Underground directly buried					X				
Under water				X[3]	X[3]				X
Near high voltage			X	X					

[1]Patch cords should be placed in a dedicated cable tray or conduit when run outside of an equipment cabinet.

[2]Fire-rated cable should always be used.

[3]Some loose tube cable manufacturers permit cable placement in shallow water. Consult the cable manufacturer for details. Otherwise, special submarine cable is necessary.

CHAPTER 5
CABLE PROCUREMENT

Before purchasing any fiber optic cable, one should give careful consideration to the lightwave equipment that will be connected to the cable and the installation environment so that the best-suited cable is selected for its particular application.

Cables are available from a standard stock, or they can be made to special order. Stock cable is usually the least expensive and has the shortest delivery time. However, the selection is limited to the most popular cable types. Special-order cables which are manufactured to the customer's specification require long lead times and are more expensive.

The following is a list of cable and fiber details that should be considered prior to an order.

Fiber Considerations

Parameter	Fiber type	Comments
1. Fiber design / specialty:	SM	NDSF, NZ-DSF, DSF, low loss, water peak suppressed, other specialty
2. Optical fiber diameter:	SM	9/125 μm
	MM	50/125, 62.5/125, 85/125, 100/140 μm
3. Operating wavelengths:	SM	1310, 1550 nm
	MM	850, 1310 nm
4. Maximum fiber attenuation:	SM	Decibels per kilometer at operating wavelengths dB/km
5. Minimum modal bandwidth:	MM	Megahertz × kilometer
6. Zero dispersion wavelength:	SM	Wavelength which has a dispersion of zero
7. Dispersion:	SM	Nanoseconds/nanometers × kilometer
8. Numerical aperture (NA):	MM	Value
9. Mode field diameter:	SM	Micrometers
10. Cutoff wavelength:	SM	Minimum usable wavelength

Fiber Considerations (Continued)

Parameter	Fiber type	Comments
11. Spectral bandwidth:	SM	Usable bandwidth for both 1310 nm and 1550 nm
12. Number of optical fibers:	SM MM	Number
13. Name of optical fiber manufacturer:	SM MM	Name

Cable Considerations

Parameter	Comments
1. Type of fiber optic cable:	Loose tube, a tight-buffered, figure 8 patch cord, pigtail, armored, dielectric, ribbon, etc.
2. Application:	Aerial, direct burial, duct installation, etc.
3. Type of cable jacket:	Outdoor, indoor, armored, etc.
4. Fire code rating:	Plenum, riser, general, unmarked
5. Dielectric or not dielectric:	Is cable a conductor?
6. Cable length:	Meters, or feet
7. Cable price:	Per unit length, bulk discount, stock discount
8. Name of cable manufacturer:	Usually not the same as optical fiber manufacturer
9. Copper pairs included:	Number and wire gauge
10. Cable outside diameter (OD):	Millimeters
11. Cable composition:	Filling compounds, strength member, armor, steel, Kevlar, etc.
12. Cable jacket:	UV rating, high-voltage rating, composition
13. Operating temperature range:	Maximum, minimum degrees
14. Installation temperature range:	Maximum, minimum degrees
15. Minimum bend radius loaded:	Centimeters
16. Minimum bend radius unloaded:	Centimeters
17. Maximum vertical rise:	Meters
18. Cable weight:	Per unit length
19. Maximum tension dynamic:	Pounds of force
20. Maximum tension static:	Pounds of force
21. Maximum cable span/sag:	Meters (for aerial installations)
22. Cable markings:	Sequential meter or foot marks, company name, cable identification, cable type, etc.
23. Crush resistance:	Pounds of force/Newtons, also impact or shotgun resistance
24. Pulling eyes:	Factory-installed, included in shipment

Cable Considerations (Continued)

Parameter	Comments
25. Cable ends:	Both cable ends accessible while on the reel without unlagging the reel
26. Optical fiber color coding:	Standard or special request
27. Cable shipping reel lengths:	Meters, feet
28. Number of cable reels:	Number
29. Type and size of cable reel:	Can installation crew handle the reel size?
30. Empty reel returnable:	Some reels can be returned for a refund
31. Cable test sheet:	OTDR test data available from manufacturer
32. Delivery and shipping dates:	Confirmed
33. Delivery location and charges:	Delivery to site
34. Cable future price:	Guarantee for additional purchase
35. Other considerations:	Other possible consideration in cable procurement

CHAPTER 6
SAFETY PRECAUTIONS

There are certain precautions that should be taken when working with fiber optics. Precautions are put into place to help maintain a safe work environment and reduce lost time due to accidents. In addition to these precautions, all other safety rules for the installation environment should also be followed. This chapter discusses safety precautions that should be observed when working with fiber optics.

Cutting and Stripping Cable. When cutting and stripping fiber optic cable, personnel should wear proper safety glasses and gloves. Tools such as cutters, strippers, cleavers, and so on can be very sharp and can cause injury.

Optical Fiber Pieces. Small pieces of cut fiber can easily fly into the air during fiber or cable cutting, fiber cleaving, or scribing procedures. Safety glasses should be worn whenever fibers will be cut, cleaved, or scribed. Pieces of fiber from the stripping or cleaving process should be disposed of immediately in a closable container labeled "Caution Scrap Glass Fiber." Cut pieces of glass fiber are very sharp and can easily damage the eye or pierce the skin and therefore should be handled with tweezers only. If a fiber splinter should lodge into the skin, it should be removed immediately with tweezers, or medical attention should be obtained.

A black, nonreflective surface should be used when cutting and splicing fiber. The black surface provides the best contrast in order to see small cut pieces of fiber.

Eating and drinking should never be allowed around a fiber work area.

Laser Light. Light from lasers used in fiber optics can be harmful to the eye. Laser light used in fiber systems is invisible; since you cannot see it, you cannot tell if the light is striking the eye and causing harm. In addition, the retina of the eye, where damage can occur, has no pain sensation. Therefore, care should be exercised when working with lasers, or fibers coupled to lasers. Never look into the end of a fiber that may have a laser coupled to it.

During normal operation of a fiber optic system when a laser is properly connected to an optical fiber, there is no risk of exposure to the laser light. The laser beam light can, however, be harmful if it becomes disconnected from the fiber.

Fiber optic systems are classified in the ANSI Z136.2 document based on the degree of the laser's optical beam's ability to damage the eye. The fiber optic laser hazard levels are divided into five Service Groups (SGs).*

- SG1 fiber system laser light is considered to be incapable of producing dangerous light levels during normal operation and maintenance, and therefore does not require any special treatment. Unnecessary exposure to laser light should be avoided.

- SG2 applies to a fiber system using visible laser light with wavelengths of 400 to 700 nm. It can cause harm to the eye if laser light is looked at directly for more than a 0.25 second.

- SG3a fiber system laser light is commonly used in fiber optic communication systems and is not normally hazardous if viewed momentarily with the naked eye. However it is hazardous if viewed through an optical instrument such as a fiber scope or microscope.

- SG3b fiber system laser light is hazardous to the eye if exposed to the light beam directly, with or without an optical instrument, or by nondiffused reflection. Total emitted laser light power from the end of the connector or fiber for this Service Group is less than +27 dBm.

- SG4 fiber system laser light is hazardous to the eye or skin if exposed to the light beam directly or by diffused reflected beam, with or without an optical instrument. This class of laser light can also cause fires. Total emitted laser light power from the end of the connector or fiber for this Service Group is greater than +27 dBm.

When working with fiber optic systems that are classified as SG1, SG2, or SG3a, ensure that the laser light does not strike the eye and do not use a fiber scope or any other optical instrument to view the end of a fiber when a laser is coupled to it.

When working with fiber optic systems classified as SG3b or SG4, the work and work area should be supervised by a qualified Laser Safety Officer (LSO); see ANSI Z136.2.

A fiber connector should never be inspected with the bare eye, fiber scope, or any other optical instrument unless it can be determined that no laser sources are connected to the other end of the fiber. Fiber scopes can be

---

*Note: The total emitted laser light power as indicated in the SGs above may vary for different fibers. When more than one wavelength is present in an optical fiber (WDM system), the Service Group emitted laser light power applies to the total optical power of all combined wavelengths.

purchased with laser filters that are more expensive than fiber scopes purchased without the filters, but the extra cost is well worth it. Additional information regarding laser safety should be obtained from the following references:

American National Standard for the Safe Use of Optical Fiber Communication Systems Utilizing Laser Diode and LED Sources; ANSI Z-136.2(1988)

American National Standard for the Safe Use of Lasers; ANSI Z-136.1 (1993)

Instructional PUB 8-1.7, Guidelines for Laser Safety and Hazard Assessment, Directorate of Technical Publications, August 19, 1991; US Department of Labor, Occupational Safety and Health Administration (OSHA, Washington DC)

Instruction CPL 2-2.20B CH-2; Chapter 17 Laser Hazards, April 19, 1993; US Department of Labor, Occupational Safety and Health Administration (OSHA, Washington DC)

Standard 29 CFR 1926.102, Eye and Face Protection; US Department of Labor, Occupational Safety and Health Administration (OSHA, Washington DC)

Standard 21 CFR and 1040.11, Laser Products and Performance Standards; US Department of Labor, Occupational Safety and Health Administration (OSHA, Washington DC)

IEC 825-2, Safety of Laser Products, Part 2: Safety of Optical Fibre Communications Systems; International Electro-technical Commission (IEC) TC76 WG5

Cable Tension. Under tension, fiber optic cable strength members can have a lot of spring in them and can easily whip back and cause injury. Care should be taken during the cable-pulling operation or whenever the strength member is under tension.

Solvents and Cleaning Solutions. A 90% or better grade isopropyl alcohol, found as a straight solution in a bottle or as premoistened wipes, is often used to clean fibers and connectors of dirt and oils. This alcohol is highly flammable and volatile. Contact with the skin or eyes can cause irritation. Inhalation of vapors should be avoided since overexposure can cause liver or kidney damage. If this alcohol is ingested, immediate medical attention should be sought.

Other liquids that are used to clean optical fibers and remove filling compound can also cause skin and eye irritation. The vapors are potentially flammable and can cause respiratory problems.

When working with these solvents, wear hand protection, keep the area well ventilated, and don't smoke or permit open flames in or near the area. Eating and drinking should never be allowed around a fiber work area.

When sold, all chemicals are required to be accompanied by a technical bulletin known as the Material Safety Data Sheet (MSDS), also known as WHMIS in Canada. This data sheet details important information regarding the ingredients in the chemical, health hazards, and special safety precautions when using it.

Fusion Splicer.   The electric spark generated by a fiber optic fusion splicer can cause an explosion in the presence of flammable vapors. A fusion splicer should never be used in a confined area such as an underground cable vault.

Always become familiar with and aware of the manufacturers' recommendations and precautions when using and installing their products.

CHAPTER 7
HANDLING FIBER OPTIC CABLE

Special precautions and handling techniques should be used whenever working with fiber optic cable, even though it may resemble copper cable. One should always keep in mind that inside the cable the optical fibers are glass and can be damaged if improperly handled.

During an installation, the two most important things to keep in mind are the following: first, the cable's minimum bend radius, and second, its pulling tension.

Minimum Bend Radius.   Fiber optic cable has a minimum bending radius, specified by the manufacturer, for loaded conditions, such as during a cable pull, and for unloaded conditions, such as the time after the cable has been installed and is in its final resting position. The cable must not be bent tighter than the loaded minimum bending radius at any time during the installation process. The unloaded bend radius is smaller and can be used only when there is no tensile load on the cable, for example, after the cable has been placed into its final resting position and is not under tension. The bending radius varies with the cable diameter and is sometimes specified as a multiple of the cable diameter (for example, 20 × OD).

Individual fibers and fiber patch cords have a smaller minimum bending radius, usually between 2 and 3 cm (see Fig. 7.1). This minimum bending radius varies with the operating wavelength and is slightly larger for larger wavelengths (for example, 1550-nm operations).

Bending the cable tighter than its minimum bend radius can damage the cable and/or increase fiber attenuation above the manufacturer's specifications. If the fiber optic cable is bent tighter than the allowed minimum bend radius or if the cable is abused, the individual fibers in the cable may have been broken even if no physical damage to the cable is evident. The abused section or the entire length of cable may need to be replaced. The cable should be tested immediately.

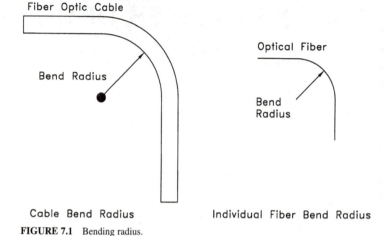

FIGURE 7.1   Bending radius.

Pulling Tension.   Fiber optic cable has a lower pulling tension than do many conventional cables. Maximum pulling tensions during installation are specified by the manufacturer and should not be exceeded at any time. The cable should be pulled by hand as much as possible. Pulling tensions should always be monitored when using mechanical pulling techniques; a strip-chart recorder is often used for this purpose. Whether by hand or mechanically, the cable should be pulled in a steady, continuous motion and never jerked. At no time should the cable be pushed. The cable should be installed using the minimum possible tension.

For permanent installed cable conditions, the tensile load on the cable should be kept to a minimum well below the manufacturer's specification for tensile load in a permanent installation. Most tensile load on a cable will occur in a vertical installation and is caused by the cable's own weight. This load should be determined and kept below the manufacturer's specifications.

General Care.   Different cable types can be handled in different manners. An outdoor loose tube armored cable can withstand much more abuse than an indoor fan-out cable. The chart below compares the ruggedness of many different fiber optic cables, beginning with the most rugged cable at the top.

Cable Type	General Characteristics
1. Outdoor loose tube armored	Most rugged; good crush resistance; can be used for direct burial installations
2. Loose tube not armored	A strong cable; may not withstand high crushing force; should be placed in a duct if it is to be buried

3. Indoor tight buffer	Care should be taken during its installation; fibers can be broken easily if sharp bends are encountered during pulling operations
4. Patch cords	Should be installed only within equipment bays and dedicated cable trays; extra care should be taken when handling and placing these cables

All fiber optic cables should be handled with care at all times. Rough handling of the cable, such as kinking, denting, abrading, or abusing, will likely break or damage the glass fibers in the cable.

Never twist the cable. If storing the cable, use a cable reel or, for short lengths, lay the cable flat in a figure 8 pattern. Ensure that the figure 8 curves are larger than the cable's minimum bending radius. To prevent crushing the cable when storing long cable lengths, support the cable crossing points in the middle of the figure 8 pattern.

Avoid deforming the cable with cable clamps, supports, fasteners, guides, and the like. All clamps and supports should have a smooth, uniform contact surface. Do not tie wrap patch cords too tightly, but tie wrap only until the cable is snug. Ensure that the cable jacket is not deformed by the tie wrap.

Cables should be placed on flat trays without sharp, protruding edges or in conduits. Avoid causing pressure points on the cable. Also avoid laying heavy objects or piling cables on top of unarmored cables. All cable bends should be smooth, with radii larger than the cable's minimum bending radius. The cable should not contact any sharp objects that could possibly cause damage to the cable.

Caution should be exercised when installing additional cables with existing fiber optic cables. Pulling eyes can have sharp edges that can easily cut through a cable jacket. Fiber optic cables should be placed in their own dedicated ducts or trays.

For direct burial installations, the cable should lie flat in a trench, free of any large stones or boulders that may deform the cable.

Avoid placing cable reels on their sides or subjecting them to shock from dropping. Do not allow vehicles to drive over a cable. Ensure that the proper cable length has been installed before cutting off excess cable. Cutting a cable too short can cause problems during splicing operations and may result in the need to replace the entire cable.

CHAPTER 8
OUTDOOR FIBER OPTIC CABLE INSTALLATION

Fiber optic cables are available for outdoor and indoor installations. The two types of cables differ in construction. The outdoor cable is quite rugged in order to withstand the harsher environment, whereas the indoor is much less rugged but more flexible so that it can be placed into confined spaces. Each type of cable, whether it is to be placed indoors or outdoors, needs different installation techniques.

The two most common outdoor installations are overhead pole line installation and underground buried cable installation. An outdoor loose tube cable is used in these applications because the loose tube design provides the fibers with the best protection from stresses that can be applied to the cable. The cable is available with a standard outdoor jacket, an extra thick jacket (often called a double jacket), or an armored jacket (normally steel).

8.1 BURIED CABLE INSTALLATION

Fiber optic cable can be buried directly underground or placed into a buried duct. Direct burial installations are common in long cross-country routes. Once proper plow equipment is set up, the installation process will proceed at a good pace. Steel-armored direct burial cables are used for these installations.

Alternatively, installing underground ducts can provide the cable with additional protection from the environment, allowing for future cable installation or removal without the need to dig. This can be beneficial in urban below-street installations. As well, a standard outdoor cable without armor, rather than heavy armored direct burial cable, can be pulled into the duct.

Before digging operations begin, soil conditions along the cable route should be investigated to determine several things: the selection of cable-placing equipment, the type of cable and duct (if used), and the installation depth. All existing underground utilities such as buried cables, pipes, and other structures along the route should be identified and located. Proper right-of-way permits must also be obtained before digging begins.

The fiber optic cable can be trenched or directly plowed under. Trenching is more time-consuming than direct plowing but allows for a more controlled installation. Trenches are dug by hand or by machine to required depths.

Plowing in a cable is faster but should be closely monitored to ensure that the cable is not damaged during the operation.

Directional boring can also be used to place cable underground. This technique is useful in situations where the surface cannot be disturbed, such as at major road, highway, or railway crossings. The directional boring machine can bore a 1- to 3-in-diameter hole, 2 to 6 ft below the surface (depending on ground conditions), for distances of over 500 m. During the boring process a duct is pulled into the hole. Once the bore is completed, the cable is pulled through the duct, bypassing the surface obstruction.

A combination of these methods—plowing in isolated areas, trenching in areas where plowing is not possible, and directional boring under roads—can also be applied.

Fiber optic cable can be placed in the depth range of 30 in to over 40 in, depending on soil conditions, surface usage, and frost conditions. For example, a deeper installation of 40 in or more may be required in a farmer's field or at a road crossing. In colder climates, the cable can be buried below the frost line to reduce the chance of cable damage due to ground frost heaves.

Trenches should be kept as straight as possible. The bottom of the trench should be flat and level with no large stones. A light backfill can be used to provide better cable load distribution, reduce possible cable damage, and decrease optical fiber microbending loss. Backfill can be slightly above original grade level to allow for later settling of the backfill (Fig. 8.1). Proper fencing should be put into place to serve as a guard around open trenches.

Bright warning tape should be buried directly above the cable to alert future digging operations. Special detectable tape or markers can be used to help locate buried cable. Aboveground cable markers can also be used, but since these may attract undesired attention to the fiber optic cable, they are less popular.

At locations such as under road crossings or rail crossings, the cable can be buried inside a 2- to 3-in-diameter-thick walled galvanized steel pipe. The pipe serves as protection and keeps the load off the cable during digging operations. Small concrete slabs (2- × 2-ft patio slabs), buried above the cable, can also be used to protect the cable.

A cable with a heavy steel armor jacket should be selected to provide crush resistance and protection from rodents. Whenever a conducting armor such as steel is used, the cable should be properly grounded at all termination points, splices, and building entrances. Typical grounding at a splice point is shown in Fig. 8.2. The cable steel armor should be connected to a ground wire using an appropriate compression-type ground clamp. The cable ground wire is then bolted to an easily accessible ground terminal strip, which is connected to one or more buried ground rods to provide a low

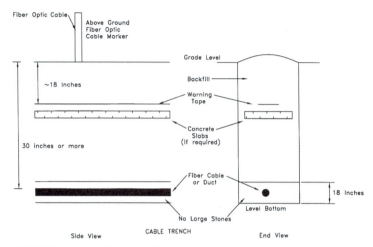

FIGURE 8.1 Buried fiber optic cable.

resistive path to earth ground. In the event a cable locate needs to be done, the cable ground wire can be unbolted and removed from the ground terminal strip and connected to the locating equipment signal generator.

Cable splices should be enclosed in a watertight splice enclosure designed for direct burial installations. The splice enclosure is then placed in a hand hole (typically a 4-ft-×-4-ft or larger sturdy box with removable lid) and buried at the same depth as the cable. The hand hole (see Fig. 8.3) is used to house excess cable slack and allow better accessibility to the splice enclosure in the event of a future re-entry.

Elaborate cabling systems that require splicing of many cables at one location may require full underground cable vault installation (manhole). The cable vault is designed to be large enough to accommodate all splice enclosures and cable entrances. Minimum bending radii for cable, duct, and innerduct should always be observed during installation.

8.2 CABLE DUCTS

Most ducts and innerducts are constructed of high-density polyethylene (HDPE), PVC, or an epoxy fiberglass compound. Ducts are usually available in black or gray. Colors identifying the innerducts as fiber optic are usually bright orange or yellow.

Inside and outside walls for ducts and innerducts are available with longitudinal or corrugated ribs. These ribs help to decrease pulling tensions

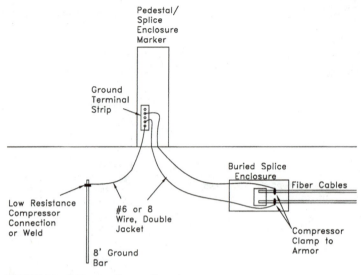

FIGURE 8.2 Cable grounding at splice points.

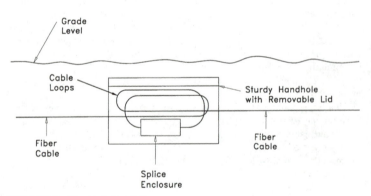

FIGURE 8.3 Hand hole installation.

during installation. The corrugated type is very flexible and can be used in locations with many turns. After a corrugated innerduct is pulled into a duct, it should be left for a day, uncut, to allow the innerduct to retreat back into the duct—with a movement resembling an accordion—through its corrugated spring action. A smooth, outside wall is available for use in tight locations.

Ducts and innerducts are available with pulling tape that has been prein-stalled by the manufacturer. This may save time during the installation process. Ducts and innerducts are also available prelubricated, which can dramatically decrease pulling tensions during installation.

Ducts and innerducts have a minimum bend radius, which means that the duct should not be bent tighter than this radius. This radius can also be spec-ified as supported and unsupported. The supported radius should be used only when the duct is bent around a supporting structure, such as in another duct or on a reel. The unsupported radius is used for duct bending in which there is no support in the bend.

Ducts provide the cable with protection and a means for future cable installation and removal. Fiber optic cable can be pulled into new or existing ducting systems. To allow for future cables to be placed in the route, either ducts can be oversized or spare ducts can be installed.

When installing cable into a public duct system, the use of an innerduct will provide the cable with protection from other companies' cable installa-tion operations (see Fig. 8.4). An innerduct provides the cable with addition-al protection from the environment and can be used in old duct installations where duct "cave-ins" are common.

After the fiber optic cable is installed into a duct or innerduct, end plugs can be installed to provide an effective water seal. The ducts and innerducts should be kept free of debris and maintained watertight at all times.

Ducts and innerducts should be sized to meet present and future cable installation requirements (see Fig. 8.5). A maximum 40% fill ratio (by cross-section area) is a good rule of thumb to follow for duct size. For example, a 0.6-in OD cable can be installed into a 1-in inside diameter (ID) duct or innerduct (sized for single cable installation only). The duct size should be increased for long installation lengths with many turns. A larger duct can help to reduce cable-pulling tensions and may also be necessary to provide room

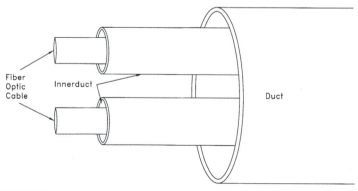

FIGURE 8.4 Ducts and innerducts.

for future cable installations. Standard duct sizes vary from 3 to 8 in ID. Innerduct sizes range from 0.75 to 2 in.

Proper thick-wall ducts should be used in buried applications and should also be capable of withstanding the crushing force of backfill and ground traffic.

8.3   DUCT LUBRICANT

High-performance fiber optic cable lubricant is used frequently for long cable duct pulls or pulls with numerous turns. The primary purpose of such lubricant is to reduce the cable's coefficient of friction, thereby lessening the tension exerted on the cable during the pulling installation procedure.

Fiber optic lubricant must have the following characteristics:

- Suitability for outdoor temperatures
- Flame-retardant properties
- Low coefficient of friction (preferably less than 0.25) when used on PE-jacketed or other types of cables
- Consistent qualities over the entire installation period
- Dry coefficient of friction, which can be noted for future cable pulls

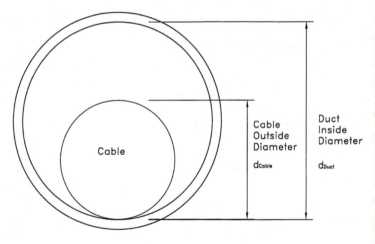

FIGURE 8.5   Duct/innerduct/conduit size.

- The inability to affect the properties of the cable jacket, conduit, duct, or innerduct during or after installation

- Testing by and approval from the appropriate authorities such as UL or CSA

Lubricant should be present at all points of the duct, and it should be applied:

- At all cable feed locations and intermediate pull locations

- Whenever possible just before bends

- With a lubricant collar and pump

Lubricant is very slippery. All lubricant spillage should be cleaned up as soon as possible using the manufacturer's recommended procedure.

The quantity of lubricant required for an installation can be roughly estimated using the following formula:

$$Q = .00378 \times L \times (NID + NOD)$$

where Q   = quantity of lubricant in liters
     L   = length of pull in meters
     NID = nominal inner diameter of duct in centimeters
     NOD = nominal outer diameter of cable in centimeters

8.4 PULLING TAPE

Pulling tape is used to pull the cable or innerduct into the buried duct. The appropriate pulling tape should be used to prevent premature breakage and damage to the cable or ducting system.

Characteristics of pulling tape should include:

- Flat type, similar to measuring tape, which reduces damage to duct during pulling operations

- Each meter (or foot) marked sequentially for easy identification of distance

- Kevlar weave for greater strength

- Design and construction to not stretch or spring

- Rating for greater-than-maximum anticipated pull tension

As mentioned earlier, pulling tape is available preinstalled in ducts or innerducts for new duct installations. Pulling tape can also be fished through or blown into a duct length.

Pulling tape is threaded through the pulling eye and sewn back onto itself to reduce the possibility of weak links. A swivel is used between the tape and cable to prevent cable twisting. Tension-sensitive, breakable links can also be used to protect the fiber optic cable from overtension.

If additional cables will be installed in the future, pulling tape should be installed with the cable and left behind for future pulls.

8.5 CABLE INSTALLATION IN DUCTS

Fiber optic cable can be pulled into existing or new underground ducting systems (see Fig. 8.6). Fiber optic cable can also be blown into a duct using an air compressor and blowing system. Duct installation is common in populated urban centers, where constant digging in streets is difficult, costly, and therefore discouraged. Once a ducting system is in place, cable installation can proceed with minimal interruptions.

Although ducts vary in size, one popular diameter is 4 in. At regular intervals, ducts are terminated in underground cable vaults (manholes). These cable vaults allow the cable route to be branched into different directions and also provide a means for accessing the ducting system.

The vaults are normally rectangular and made from concrete. When a cable ducting system is created, several ducts are installed to provide sufficient

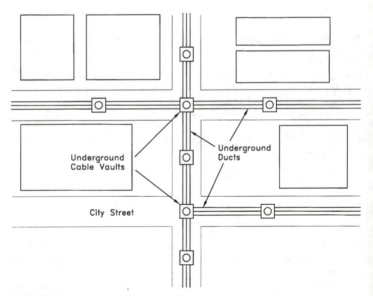

FIGURE 8.6 Underground city ducting system.

cable capacity for present and future requirements. The ducts terminate on the cable vault wall in an ordered manner, as shown in Fig. 8.7.

Installation by Cable Pull. Fiber optic cable can be pulled into an existing or new duct. The longest possible length of uninterrupted cable is pulled to reduce the number of splices. Cable lengths are determined by engineering design, considering such factors as:

- Pulling tension
- Route length
- Number of turns
- Pull direction
- Splice enclosure location
- Accessibility
- Fiber link attenuation

Before the start of any cable pull, all ducts and cable vaults should be carefully inspected for damage or deterioration, and to address any safety concerns. To ensure a safe working environment, caution should always be exercised before entering any cable vault.
Remember the following:

- Open underground vaults should be clearly marked, guarded, and enclosed within barricades.
- Tests should be performed for dangerous gases. All vaults should be properly ventilated.
- No open flames should be permitted in or around the vault.

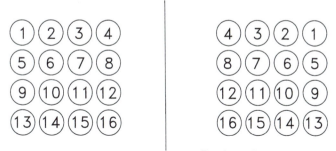

FIGURE 8.7 Typical duct wall configuration and assignments.

- Ladders, pulling irons, rails, and the like should be checked for corrosion and to ensure that they are securely fastened.
- Vehicles should be parked in such a way that their exhaust gas does not enter the vault.
- Care should be taken so that no existing cables are damaged.
- When pumping water out of an underground vault, caution should be exercised. If gasoline or oil is found, the fire department should be contacted.
- Traffic conditions should be assessed prior to entering the underground vault, and the proper permits should be obtained from city officials.
- The electric spark generated by a fusion splicer can cause an explosion when flammable gases are present; fusion splicers should therefore not be used in vaults.
- If vaults contain live high-voltage lines, the local utility or company that owns the vault should be called to determine their location and assess possible dangers.
- Safety and health regulations for industrial establishments and construction projects should always be observed.

General Cable Pull Method. Perform all prescribed safety procedures and proceed as follows:

1. Open all underground cable vaults and ensure that they are safe and clear.
2. Identify all ducts to be used for the cable placement.
3. Make sure that all ducts are clear. Duct cleaning may be required.
4. If cables are present in ducts through which the fiber optic cable is to be pulled, the existing cable type should be identified and the owner of the cable called to inform him or her of the action, and to identify any safety concerns.
5. To minimize cable tensions, reel vault locations should be set near the sharpest bend locations. Pulling and reel locations should also be set at corner vaults where possible. (See Fig. 8.8.)
6. When a cable is pulled around a bend, remember that there is added pressure against the duct wall and crushing pressure on the cable.
7. Identify the pulling and reel underground vaults. Set up all equipment appropriately (see example in Fig. 8.9).
   - The cable-pulling winch should be equipped with a strip-chart tension recorder and dynamic electronic display; winch speed should be variably controlled.
   - A lubricating collar and feeder tube should be installed in the feeding vault.
   - A proper-sized sheave should be installed in the pulling vault.

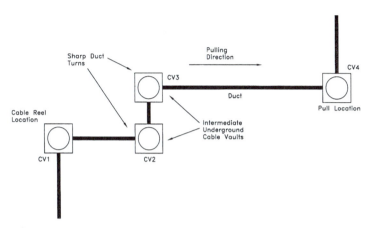

FIGURE 8.8   Reel and pull location.

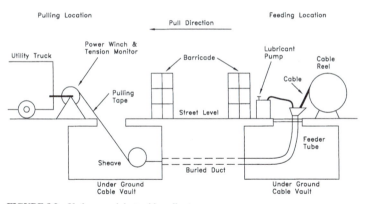

FIGURE 8.9   Underground duct cable pull setup.

8. A continuous length of pull tape should be installed in the duct route.

9. Before pulling cable and while the cable is still on the reel, all optical fibers in the cable should be tested with an OTDR and bare fiber adapter to ensure that they are acceptable.

10. If cable is to be placed in innerducts, install the innerducts first:

- Position the reel of the innerduct so that no obstructions prevent it from being pulled into the vault.
- Attach the proper pulling eye, with swivel, to the innerduct and attach the pulling tape to the eye.

- Ensure that all sheaves and capstans have the proper radius for the innerduct.
- During pulling operations, no personnel should be in the vaults. Extreme care should be exercised during the pull so as not to catch loose clothing, hands, or other objects in moving machinery. Keep loose clothing away from all moving parts. All personnel along the cable route should be in continual radio contact with each other.
- Pull as much of the innerduct by hand as possible. Apply lubricant as required. Use shorter lengths, where required, and then connect the lengths with proper innerduct connectors. Connect pulling tapes as well.
- Where hand pulls are not possible, pull the innerduct with a winch. Continuous tension monitoring is not necessary, but the maximum tensions for the innerduct should not be exceeded.
- At corners or bends, sheaves may be needed to provide the innerduct with proper support. Ensure that sheaves have the proper diameter to accommodate the cable and the innerduct's minimum bend radius.
- All lengths of innerduct and pulling tape are connected to provide one continuous length for the cable pull.
- If a corrugated innerduct is used, do not cut it to length immediately. After the pull is complete, allow the innerduct sufficient time to regain its normal shape. This will allow enough slack for the innerduct to retreat back into the duct.
- Sufficient innerduct length should be left for mounting and to allow for contraction and expansion (see the innerduct's specifications).

11. Properly attach the pulling eye and swivel to the cable. Ensure that the pulling eye and swivel assembly can easily fit through all the ducts and innerducts.

12. Do not use a woven cable grip in place of a pulling eye (except where specifications call for this, usually for short-length hand pulls).

13. Attach the installed pull tape to the swivel.

14. Adjust the capstan and sheave radius if required.

15. Pull the cable as much as possible by hand. For many cables, tension monitoring on hand pulls is not required.

16. Add lubricant generously to the cable feeder and to any midlubricant positions.

17. If hand pulling is too difficult, pull the cable at low speed with the winch. Avoid jerking motions. Keep pulling tensions well below the minimum cable-pulling tension. Continually monitor and record cable tensions on a strip chart. Make sure the cable is fed off the reel without twisting.

18. Turn the cable reel by hand to maintain slack between the reel and the feeding collar.

19. Avoid start-and-stop actions since greater tensions are required to start a cable from rest than to keep it in motion.

20. During the pull, if cable tension is near the maximum allowed, stop the pull and check the cable route for obstruction. Low lubricant levels or other difficulties may cause a high-tension problem.

21. After the cable tension problem is corrected, restart the pull and closely monitor the cable tension. If the tension is still high, stop the pull. The following measures can be performed to remedy the problem:

- Inspect each bend closely. Ensure that each is smooth with no obstructions or sharp corners. Increase the bending radius if possible. Check to see if all sheaves turn easily.
- Ensure that the feeding cable reel turns freely, with no added tension to the cable.
- Try adding lubricant with a midlubricant pump before each bend.
- Shorten the pulling route length. Move the pulling location to the route's midpoint and restart the pull. Pull the entire length of cable out through this vault and coil it on the ground. Using the figure 8 coil pattern to reduce cable twisting, move the pull location back to the original position and pull the cable through the rest of the cable route.
- Alternatively, use a second pulling winch at an intermediate vault location to assist in the pull. Pull out enough fiber optic cable to wrap about one to three turns around the second winch's capstan. This may introduce one to three twists into the cable.
- Install sheaves in the intermediate vault as required.
- By radio, coordinate both pulling winch operations.
- Ensure that there are always 3 m or more (10 ft or more) of cable slack coming off the intermediate winch before reentering the vault.
- The intermediate winch should pull the cable simultaneously with the main winch while maintaining a continuous slack loop.
- Closely monitor the cable tension at the intermediate winch.
- If maximum pull tension is recorded again, move the pull location closer to the feed vault until a successful pull can be completed.

22. At intermediate vaults, when the pulling is stopped, observe the cable for sufficient lubrication.

23. Continue the pull until a sufficient length of cable (necessary for splicing outside of the vault) is pulled into the pulling vault. Additional cable may need to be pulled so that there is enough cable to be pulled back for racking in each intermediate vault.

24. Depending on engineering design, at least 6 m (20 ft) of excess cable at each end must be left coiled for future splicing and racking and for emergency repair.

25. After a sufficient length of cable has been measured at the end, cut the cable from the reel (ensure that proper authorization is received).

26. Test the cable with an OTDR to make sure it has not been damaged in the pulling process.

27. After the cable has been successfully tested, cable racking can begin. If an innerduct is used, it is best to rack it in the vault without cutting it. If the innerduct is too stiff to be racked properly, then, if absolutely necessary for proper mounting, slit and remove the innerduct using the proper tools. Be extremely careful not to nick, cut, or damage the fiber optic cable inside. Rack the cable as required or clip it to the vault walls or ceiling. Place the cable as high as possible in the vault. Cable slack may need to be pulled back from the ends for racking. A corrugated slit duct can be placed around the cable for additional protection.

28. Before the innerduct is cut, ensure that an adequate amount is left over to allow for innerduct expansion. Innerduct end plugs can be used to seal the innerduct from debris and water seepage.

29. Fiber optic cable tags should be placed on the cable in each vault to identify the cable, the owner, and the owner's telephone number (see Fig. 8.10).

FIGURE 8.10   Fiber optic cable tag.

Installation by Cable Blowing. Fiber cable can also be installed into a duct or innerduct by cable-blowing technique. Specialized cable-blowing equipment is connected to one end of a duct or innerduct. The other end is left open. The cable is fed through the blowing equipment and pulled into the duct for about 200 ft. Then the compressor is turned on and duct pressure is increased to 100 psi (depending on equipment). The rushing air in the duct at this pressure pushes the cable through the duct. Cable can be installed quite quickly using this technique and with no damage to the cable. Cable installation speeds of 200 ft per minute are common. The installed duct should be smooth-wall HDPE sized for proper fill ratio (50 to 80%). Higher cable to duct fill ratios can achieve longer installation distances. This is because the cable can be pushed through the duct without forming waves in the duct. However, the higher fill ratio can hinder cable movement through duct bends. A continuous duct length should be used, or a number of shorter lengths can be connected. If connecting ducts, ensure that the connection is properly sealed, airtight, and strong enough to withstand the high air pressure inside the duct during the blowing process.

Blowing equipment instructions should be followed, and trained installation crews should be used for this process. Installation lengths vary, depending on the number of turns in the duct, fill ratio, and the type of blowing equipment, but can exceed 2000 ft.

8.6 AERIAL INSTALLATION

Aerial installation can be performed by lashing the fiber optic cable to an existing steel messenger or by installing a self-supporting fiber optic cable along a pole span (see Fig. 8.11).

Extreme caution should be observed when performing an aerial installation. The proper personnel should be contacted so that they are on the site when work is performed near high-voltage lines. Safety guidelines for aerial installation are:

- Cables should not be installed in wet conditions.

- Cables that are installed in the vicinity of high-voltage power lines should be grounded, including all-dielectric cables.

- Maintain proper clearance between the fiber optic cable and power cable at all times. Always make allowances for power cable sag due to weather and current conditions. Cable sag increases in warm weather or when power cable is passing heavy current, and overhead power lines without insulation are common.

- Make sure that all personnel are properly trained for pole line work.

- Ensure that fiber optic cable meets or exceeds electric field radiation specifications when it is placed near or on high-voltage power lines.

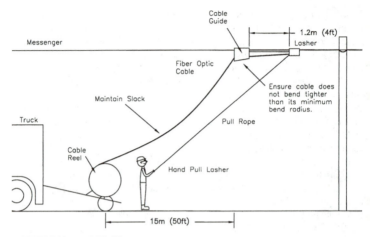

FIGURE 8.11   Aerial lashing setup.

A steel messenger is installed between the poles at the designed tension and sag to support the fiber optic cable. The messenger should always be properly grounded. Avoid zigzagging the messenger from one pole side to the other and instead, make sure it is kept on one side as much as possible.

A cable reel trailer and truck are used to dispense the fiber optic cable onto the messenger. The fiber optic cable should be properly grounded before beginning any installation. A cable guide and cable lasher are used to secure the cable to the messenger. An aerial bucket truck should follow the lasher to ensure it is operating correctly and adjusted properly at pole locations.

At locations where the fiber optic cable is to be placed onto a messenger with other existing cables, make sure that the lashing machine can accommodate other cables. If not, the cable can be lashed or tie wrapped manually from a bucket truck.

The cable can also be pulled onto the messenger and poles. This requires wheel blocks to be placed on each pole and along the span. The pulling rope is threaded through the blocks and attached to the cable. A winch with a tension monitor then pulls the fiber optic cable from the reel and onto the pole span. This method can be used when crossing highways, rivers, and rough terrain.

Once the cable is in place, it is lashed to the messenger. For self-supporting cable, the cable sag (tension) is adjusted to engineering specifications and is then secured to the pole and dead-end clamps.

Fiber optic cables can enter a building either by aerial or underground routes. Factors such as the number of cables, appearance, bending radius, inside cable route, and security should be considered before the cables are installed.

At each pole location, the cable is formed to an expansion loop to allow for expansion of the messenger (see Fig. 8.12). Because of the optical fiber's glass properties, fiber optic cable expands or contracts very little with changes in temperature. Therefore, to reduce tension on the fiber optic cable when it is strapped to a steel messenger, a small expansion loop is added.

The expansion loop's size (the amount required to compensate for expansion) is determined by engineering design. The fiber optic cable's bending radius should be observed here. The length of D of the loop in Fig. 8.12 should be greater than twice its depth R. The length D should also be greater than twice the cable's minimum bending radius.

A bright spiral identifier cover can be placed around the expansion loop portion of the cable to identify it. A fiber optic cable tag can also be added to identify the cable.

Some new fiber optic cables available today do not require expansion loops as long as they are installed to the manufacturer's specifications. Self-supporting fiber optic cables also do not require expansion loops.

At terminating poles, where the cable is routed underground, a metallic conduit is used as a pole riser to protect the cable from damage. This conduit should be properly grounded. Depending on engineering design, at least 6 m (20 ft) of excess cable at each end is left coiled for future splicing. If splicing is required at midroute, the weatherproof splice enclosure and the excess cable should be installed on the messenger shown in Figs. 22.2 and 22.3.

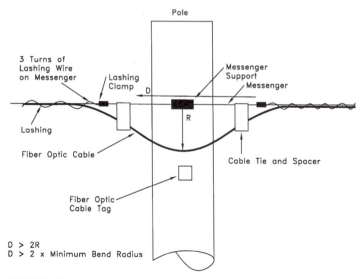

FIGURE 8.12 Expansion loop.

General Aerial Installation Procedure

1. Ensure that all safety precautions are observed.

2. Install the messenger to the proper sag tension, ensuring that it is properly grounded.

3. Prepare the equipment as shown in Fig. 8.11. Install the lasher and cable guide on the messenger. The cable guide should be kept 4 ft ahead of the lasher with a stiff rod. Ensure that the cable guide's trough bend is smooth and is larger than the cable's minimum installation radius. A sheave with the proper radius can also be used.

4. Raise the cable to the cable guide and into the lasher. Keep the cable reel at least 15 m (50 ft) away from the lasher. Ensure that the cable is not bent tighter than its minimum bending radius.

5. Install the lashing and secure it to the messenger with the lashing clamp.

6. To temporarily hold the cable in place on the messenger, tie wrap the cable to the messenger at the lashing clamp.

7. Adjust the lasher for proper operation.

8. Attach a pull rope to the lasher. The lasher's pull rope should be hand pulled.

9. Begin the hand-pulling operation by pulling the lasher at a constant speed and driving the vehicle carrying the reel so that it is 50 ft from the lasher. Maintain a slight tension on the cable and keep the cable reel underneath the messenger at all times. Make sure that the proper cable bend radius is maintained at all times. Do not allow the fiber optic cable to wrap around the messenger.

10. Each time a pole is reached, the pulling should stop. The lasher and guide are then disconnected and moved past the pole. The lashing wire is terminated with a lashing clamp, and the cable is formed into an expansion loop (if required).

11. Once the lasher and guide are set on the other side of the pole and the expansion loop is complete (as shown in Fig. 8.12), the lashing operation is continued.

12. Install fiber optic cable warning tags where required.

CHAPTER 9
INDOOR CABLE INSTALLATION

Fiber optic cable placement in buildings should be carefully planned. Building floor plans can greatly aid with the route planning and should be marked appropriately with the route for construction and kept for future reference. Existing building passages, risers, conduits, and cable trays should be taken advantage of wherever possible.

Cable installation must meet or exceed the National Electric Code (NEC) and local building codes. NEC indicates that nonconductive, proper fire-rated fiber optic cables can occupy the same cable tray or raceway with conductors for electric light, power, or Class 1 circuits operating at 600 V or less. Also, fiber optic cables can be placed in the same raceway, cable tray, or enclosure with cable-TV, telephone, communication circuits, and Class 2 and 3 remote control signaling circuits. However, these fiber optic cables cannot be placed in the same enclosure that houses electrical terminations. Refer to the NEC for details.

9.1 CONDUITS AND CABLE TRAYS

Conduits and cable trays must meet mechanical restrictions imposed by the fiber optic cable. The critical constraint is the bending radius. All bends must have smooth curves. If a cable is to be pulled into a conduit or cable tray, the conduit's bending radius must be larger than the cable's minimum bending radius for loaded conditions. If the cable is laid into a tray and will not be pulled, then the tray's bend can be as tight as the cable's unloaded minimum bending radius (see Chap. 7).

If other cables are to be piled onto the fiber optic cable, an armored cable with high crush resistance or an added duct should be used for greater protection. This added protection will help reduce potential damage due to pressure points on the cable. All conduit and cable-tray fittings should be selected carefully to ensure that sharp edges or bends do not touch the cable at any time.

Conduits should be sized to meet present and future cable installation requirements. A 53% fill ratio (by cross-sectional area) is a good rule to follow for a minimal conduit diameter for one cable (see Fig. 8.5). National Electric Standard (NEC) for conduit fill limits are: 53% for one cable, 31% for two cables, and 40% for three or more cables

Fill ratio of 1 or more cables

$$= \frac{d^2 cable_1 + d^2 cable_2 + d^2 cable_3 \ldots}{D^2 \ conduit} \times 100.$$

For example, a 0.6-in OD cable can be installed into a 1-in ID conduit (sized for single cable installation only). A larger-diameter conduit should be used to reduce pulling tensions for longer routes or for installations with many turns.

Use only conduits approved to all fire and building codes. When routing conduits through fire walls, ensure that the properly rated fire wall sealant is used after the installation. Conduits should also be sealed to prevent moisture, dust, or smoke migration.

A note of caution: Before drilling through any floors, walls, or ceilings, make sure that the drill will not come into contact with or cut any embedded tension cables, electrical wiring, plumbing, conduits, or other objects. For masonry and other construction, X-rays of the drill area may be needed to verify the presence of a clear, unobstructed drill path.

9.2   PULL BOXES

Pull boxes are used to break up long conduit lengths for easier and lower tension pulls. Generally, they are placed near bends and in long straight spans. Install at least one pull box after every second 90° bend and in long conduit spans. A straight-through pull box length should be at least equal to four times the cable's minimum bend radius (see Fig. 9.1). A corner pull box length should be at least three times the cable's bending radius and one radius deep to the conduit (see Fig. 9.2). When pulling cable out of pull boxes, ensure that the cable's minimum bending radius is not compromised. Cable pulled through a corner pull box should first be pulled out into a loop (see Fig. 9.2). The sharp corner of the pull box can easily damage the cable and fibers.

9.3   VERTICAL INSTALLATIONS

In vertical installations, the weight of the cable creates a tensile load on itself. This load should not exceed the cable's maximum tensile load for a

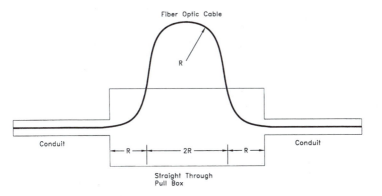

FIGURE 9.1 Straight-through pull box.

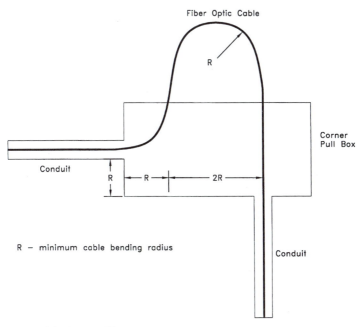

FIGURE 9.2 Corner pull box.

permanently installed position. A maximum cable vertical rise is also specified for a cable and should not be exceeded. Both cable load and rise conditions should be kept well below the manufacturer's specifications. Clamping a vertical cable to a support at intermediate points can reduce cable tensile loading. The clamping force should be no more than is required to prevent the cable from slipping. The cable should not deform in any way. Cable clamps should have a smooth, uniform surface to prevent cable damage.

If frequent clamping is not possible, in order to suspend the cable, cable hangers can be used at the top of the vertical rises and at intermediate locations along the vertical rises. Cable hangers that will not damage or deform the cable should be selected. A popular choice for such an installation is the mesh grip or split mesh grip hanger as shown in Fig. 9.3.

Once the mesh grip is placed onto the cable, it can be tightened to provide an effective slip-free support. It is then mounted onto a wall hanger. Make sure that the cable bend is greater than the cable's minimum bending radius at the top of the cable rise.

Tight-buffered cable is commonly used in vertical installations because of its specified high vertical rise capacity. All indoor cables should meet or exceed fire code regulations.

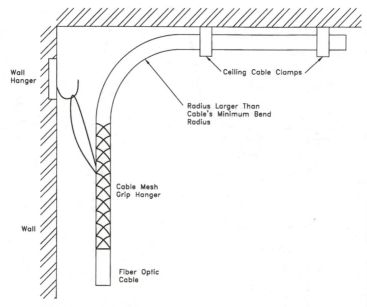

FIGURE 9.3   Cable mesh grip hanger.

9.4 BUILDING ROUTES

Outdoor cables can be very stiff and heavy and difficult to install in tight building passages. When an outdoor fiber optic cable enters a building, it should be spliced to an indoor-type fiber optic cable within 50 ft from the cable entrance (to meet NEC code). A splice enclosure or patch panel that can handle a number of cables for distribution can be used at this point. This provides a common optical fiber distribution point for the building and allows the proper indoor fire-rated cables to be used throughout the building. It also provides a reel or pulling location for installing the outdoor cable. The splice enclosure, patch panel, conduit, and cable should all be properly grounded.

As Fig. 9.4 shows, the indoor cable runs from the splice enclosure to the equipment patch panel on an upper floor. Normally, it is placed into a fire-rated conduit or tray for the entire route. The conduit can be sized (as shown in Fig. 8.5) or oversized, to accommodate additional future cables. In some instances, the use of conduits can be prohibitive, so a fire-rated armored cable may be used instead.

Horizontal cable routes can be placed above suspended ceilings or under

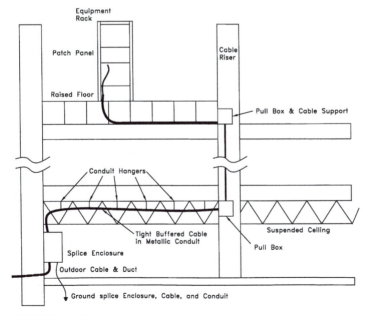

FIGURE 9.4 Cable route example.

raised floors. The conduit, cable tray, or cable is secured to the roof or floor with proper clamps and labeled "Fiber Optic Cable" (see Fig. 8.10).

Cable vertical rises in tall buildings are made in riser closets. When riser closets are not available, holes are drilled through the floor to route the cable or conduit. At the lightwave equipment rack, the cable can enter through the rack's top or bottom. Loops of excess fiber optic cable can be left beneath or above the rack to allow for future rack or patch panel moves or for resplicing.

Proper authorization should be given before cutting any excess fiber optic cable. At least 6 m (20 ft) of excess cable at each end should remain for future splicing or termination. An additional length can be left uncut if required.

9.5 GENERAL CABLE INSTALLATION PROCEDURE

Pulling fiber optic cable indoors is usually done by hand. Lubricant can be added in difficult pulls or when cable is placed with other existing cables.

Method.   Perform all prescribed safety procedures and proceed as follows:

1. Identify and open all pull boxes, conduits, and cable trays and ensure that they are unobstructed and meet fiber optic cable requirements (see Secs. 9.1 and 9.2).

2. If fiber optic cable is to be placed into conduits or cable trays that contain cables, the existing cables should be identified, and the owner of the cables should be informed of the installation. All safety concerns should be identified.

3. A continuous length of pull tape is installed in the complete conduit or tray route.

4. Before proceeding with the installation, the fiber optic cable should be tested to ensure that it is acceptable (see Chap. 15).

5. Properly attach the pulling eye and swivel to the cable. Ensure that the pulling eye and swivel assembly have no sharp edges and can easily fit through all conduits, pull boxes, and cable trays.

6. Attach the installed pull tape to the swivel.

7. Hand pull the cable through the first section of conduit or tray and out of the first pull box. Make sure that cable bends are larger than the cable's minimum bending radius at all times. Add lubricant if required, avoid any jerking motions, and do not push the cable at any time.

8. Coil the cable on the floor in a figure 8 pattern to avoid twisting. Continue the cable pull until all the cable has been pulled through the first section.

9. Feed the cable back into the pull box, and pull the cable through to the next pull box. Continue this procedure until all the cable has been installed. When feeding cable back into the pull box, make sure that the cable's minimum bending radius is not compromised.

10. If open cable trays are encountered, the cable can be laid into the tray instead of being pulled.

11. Continue this process (steps 1 to 10) until all the cable has been completely installed into the conduit or tray route.

12. Depending on the engineering design, at least 6 m (20 ft) of excess cable at each end should be left coiled for future splicing and racking.

13. Finally, the fiber optic cable should be tested to ensure that it has not been damaged during the installation (see Chap. 14).

CHAPTER 10

FIBER OPTIC CABLE GENERAL INSTALLATION GUIDELINE

The following general guidelines can be used to help many fiber optic cable installations:

1. Identify the exact fiber optic cable route and ensure that it meets all installation specifications. Obtain all required fiber optic cable installation authorizations and permits along the route.

2. Complete the fiber optic cable engineering design, including the following:

   • Identify lightwave equipment that will be used and ensure that equipment will function properly with the fiber cable design.
   • Determine fiber type and calculate or measure (for existing cable) total optical loss, dispersion, and other bandwidth-limiting concerns.
   • Determine cable type—outside plant, inside plant, loose tube, tight buffered, armored, fire rating, number of fibers, etc.
   • Determine splice locations, cable reel lengths, splicing crews, and fiber patch panel locations.
   • Establish the optical fiber's connector type and connector installation procedure (pigtail, breakout cable, direct fiber connection).
   • See Chap. 5 for additional details.

3. Identify all installation hazards and safety concerns.

4. Ensure that proper installation and test equipment is available. Make sure that all personnel are properly trained to handle and install fiber optic cable. Determine the correct installation procedure and schedule.

5. Order fiber optic cable(s) and all equipment required, as per design, allowing for adequate delivery time.

6. Once cable(s) is (are) received, perform the fiber optic cable reel test prior to cable installation (see Sec. 15.1—Fiber Optic Cable Tests).

7. Prepare the fiber optic cable route and install all conduits, ducts, innerducts, messenger cables, and so on, as required.

8. Install fiber optic cable as required according to engineering design.

9. Splice all separate cable lengths together, as required. OTDR testing during splicing may be required.

10. Terminate fiber optic cable in appropriate patch panels or splice enclosures and complete cable installation.

11. OTDR and power meter test complete fiber optic facility (see Sec. 15.1). Ensure that fiber facility test results fall within design criteria.

12. Install all terminating lightwave equipment, modems, multiplexers, and so on.

13. Connect lightwave equipment to fiber optic facility commission and test (BERT or equivalent).

14. Record all required fiber optic facility details, including proper route drawings, fiber assignments, loss readings, OTDR traces, etc. (see Chap. 23).

15. Prepare emergency repair plans (see Chaps. 21 and 22).

CHAPTER 11
SPLICING AND TERMINATION

11.1 SPLICE ENCLOSURES

Splice enclosures are used to protect stripped fiber optic cable and fiber optic splices from the environment, and they are available for indoor as well as outdoor mounting. The outdoor type should be weatherproof, with a watertight seal.

Figure 11.1 shows a typical wall-mounted splice enclosure. The fiber optic cable is supported by cable ties, and the strength member is securely fastened to the enclosure's support. Metallic strength members are grounded.

The cable sheath stops at the splice enclosure's cable ties. Optical fiber tubes, individual thick-buffered fibers, or pigtails are supported by the tube brackets and continue on to the splicing trays. Individual optical fibers should not be exposed. The actual splices are contained in the splice trays.

11.2 SPLICE TRAYS

Splice trays (Fig. 11.2) are used to hold and protect individual fusion or mechanical splices. They are available for many types of splices, including various brand-name mechanical splices, bare fusion splices, heat-shrink fusion splices, and so on. The splice tray should be matched to the type of splice used. For example, a splice tray designed to house a mechanical splice will not hold a bare fusion splice.

Splice trays can be optical wavelength–sensitive. A splice tray designed for only 810-nm wavelengths may cause optical loss if 1550-nm wavelengths

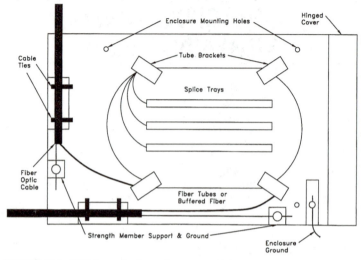

FIGURE 11.1 Wall-mounted splice enclosure.

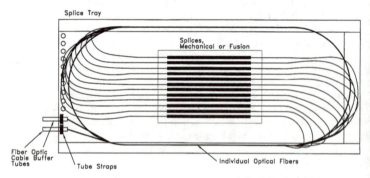

FIGURE 11.2 Splice tray.

are used. The splice tray's operating wavelengths should confirmed with the manufacturer's specifications.

Splice trays normally hold up to 12 splices, and several trays are used together to splice a large fiber cable. Each tray provides space for mounting fiber splice protectors and excess fiber. Fiber buffer tubes should enter the splice tray at one end only. At this end the buffer tubes stop and are secured to the tray where the individual fibers are exposed. If some fibers need to be routed to a different tray, proper buffer tube splitters should be used. Unprotected fibers should not be exposed outside the splice tray. Care should be exercised when mounting the splice in the tray. The individual fiber bending radius should be kept as large as possible (greater than the minimum fiber bending radius).

11.3 PATCH PANELS

A fiber optic patch panel (also called fiber distribution panel) terminates the fiber optic cable and provides connection access to the cable's individual fibers (see Fig. 11.3). Using fiber patch cords, individual cable fibers can be cross-connected, connected to lightwave equipment, or tested at the patch panel. It also allows for labeling of the cable's individual fibers and a link demarcation point.

The patch panel is designed with two compartments: one which contains the bulkhead receptacles or adapters, and the second which is used for splice tray and excess fiber storage. Patch cord management trays are optional for some patch panels and make possible the neat storage of excessive patch cord lengths.

The patch panel's bulkhead panel contains the adapter (also known as the receptacle and barrow). The adapter allows the cable's fiber connector to mate with the appropriate patch cord connector. It provides a low optical loss connection over many connector matings (see Fig. 11.4).

Patch panels are available in rack-mounted or wall-mounted styles and are usually placed near terminating equipment (within patch cord reach).

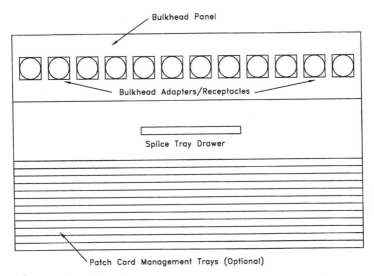

FIGURE 11.3   Patch panel.

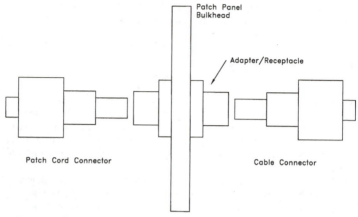

FIGURE 11.4   Bulkhead and adapter.

If mounted in a rack, vertical location should be considered. Adequate space above or below the panel should be provided for the fiber optic cable(s) entering the enclosure. Equipment mounting, in this area, may not be possible because of the cables. The fiber optic cable's minimum bending radius should always be observed when terminating cable at patch panels.

Fiber optic cable can be terminated in a patch panel using both pigtail or field-installable connector fiber termination techniques (see Fig. 11.5). The pigtail technique requires that a splice be made and a splice tray be used in the patch panel. However, this technique can provide for the best quality connection and can be the quickest to complete.

The field-installable connector technique usually takes longer than splicing but does not require a splice or splice tray to be mounted in the patch panel. This reduces material costs and allows for a smaller patch panel to be used (i.e., saving on rack space). Also, the connector loss and quality may not be as good as factory-purchased pigtails. The technique is often used when terminating a tight-buffered cable.

Before making any fiber optic connection, ensure that all connectors and receptacles are clean. Connections should only be "finger tight." Never rotate the connector ferrule during connection.

11.4   SPLICING

Optical fiber splicing is the technique used to permanently join two optical fibers in a low-loss connection. This connection can be made using one of two methods—fusion splicing or mechanical splicing.

Fusion splicing provides the lowest-loss connection. This technique uses a device called a fusion splicer to perform the fiber splicing. The fusion splicer precisely aligns the two fibers and then generates a small electric arc to bond the two fibers together. A good fusion splicer will consistently provide low-loss splices, usually less than 0.1 dB for single-mode or multimode fibers. However, such splicers are quite expensive and bulky and can be difficult to operate. For proficiency in an operation, proper training is required.

A note of caution: The electric spark generated by a fusion splicer can cause an explosion when flammable gases are present. Fusion splicers should not be used in vaults or confined spaces that may contain flammable gases.

Mechanical splicing is an alternate splicing technique that does not require a fusion splicer. It uses a small, mechanical splice, approximately 6 cm long and 1 cm in diameter, that permanently joins the two optical fibers. A mechanical splice is a small fiber connector that precisely aligns two bare fibers and then secures them mechanically. A snap-type cover, an adhesive cover, or both, are used to permanently fasten the splice. They are small, quite easy to use, and are very handy for either quick repairs or permanent installations. They are available in permanent and reenterable types.

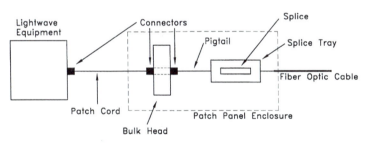

a.

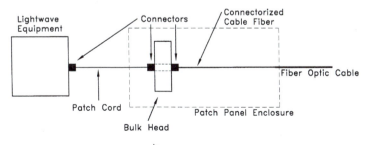

b.

FIGURE 11.5   Patch panel cable termination.

Mechanical splices are available for single-mode or multimode fiber, but their connection losses are greater than in fusion splicing and can range between 0.1 and 0.8 dB (although usually less than 0.5 dB).

Splicing a fiber optic cable is accomplished by stripping the cable jacket with a jacket stripping tool to expose the fiber tubes or individual buffered fibers. The tube or buffer is then stripped to expose the individual coated fibers, and any gel on the tube or fibers should be carefully cleaned off.

In order to expose the fiber cladding, the coating is stripped a short length at a time (1 cm or less). The fiber is then cleaved and spliced. The stripping process uses hand tools that are specially designed not to nick or damage the individual fibers.

Cleaving a fiber provides a uniform, perpendicular surface that will allow maximum light transmission to the other fiber. High attenuation at a splice is usually due to either a bad splice or a bad fiber cleave. A good-quality cleaving tool is recommended.

For a fusion splice, a splice protector is immediately placed covering the fiber splice. The splice protector adds physical strength to the fiber splice and protects it from contaminants.

After a splice is completed, with either the fusion or mechanical splice technique, the spliced fiber is stored in a splice tray. The splice protector for fusion splices, or the mechanical splice, snaps into a splice holder found in the center of the splice tray. If a splice holder is not there, or if it is not the proper size or type, then the splice can be glued to the tray with a dab of contact cement. Do not use tape to hold the splices to the splice tray.

The splice tray is then mounted into the splice enclosure. This protects the splice and exposed fibers and provides a means for accessing a splice if required. For small-fiber-count cables, and when using tight-buffered fiber optic cable, connectors can be installed directly onto the cable's optical fibers without the need for splicing or the use of splice trays and enclosures.

Outdoor splicing of cable is commonly done in a splice van or tent to protect the exposed fiber and equipment from the environment. It also prevents tools, equipment, and fibers from being blown around in the wind.

During cable installation, it is important to leave sufficient slack for the fiber optic cable and splice enclosure to be brought inside the van or tent for splicing. All splicing should be done on a large, clean table with plenty of room for all splice equipment and cables. A splicing team typically consists of two technicians: a splicer and an OTDR or optical power meter operator.

Splicing Tool Kit

Ruler

Alcohol cleaning solution

Cable gel cleaning solution

Cotton swabs, no-residue

Paper tissues, no-residue

Fiber cleaver

Fiber coating stripper

Fiber buffer and tube stripper

Fiber jacket stripper (specify jacket size)

Utility knife

Diagonal cutters (used to cut cable or fibers to proper length)

Scissors, serrated blade

Emery cloth (used to securely hold individual fiber between fingers during coating stripping)

Tweezers (used to handle cut fiber pieces)

Closable container (used to dispose of cut fiber pieces)

Protective gloves (to guard hands from cleaning solutions)

Splice protectors (for fusion splices)

Mechanical splices or fusion splicer

Splice tray and patch panel or enclosure

OTDR or power meter and light source

Large table and chairs

Black paper tablecloth (provides best contrast for seeing individual fibers)

Splicing Method

1. Identify all optical fibers to be spliced and splicing assignments. Plan the exact fiber route in the splice enclosure, from the cable entrance into the splice tray. Ensure that the splice tray is equipped with the proper splice holders: mechanical, bare fusion, heat shrink, and so on.

2. Strip back about 3 m of the cable jacket to expose the fiber tubes or buffered fibers (the exact length will vary for different types of splice enclosures). Use cable rip cord to cut through the jacket. Then carefully peel back the jacket and expose the insides. If rip cord is not available, carefully strip the jacket with a jacket stripping tool or utility knife, and make sure that the optical fiber tubes or buffers are not damaged. Cut off the excess jacket. Clean off all cable gel from exposed tubes or buffers with cable gel remover. Use gloves to protect your hands from the cleaning solution. Separate the tubes and buffers by carefully cutting away any yarn or sheath. Leave enough of the strength member to properly secure the cable in the splice enclosure (see Fig. 11.6).

3. For a loose tube cable, strip away about 2 m of fiber tube using a buffer tube stripper and expose the individual fibers. The exact stripping length will vary for different splicing techniques and splice trays. For a tight-

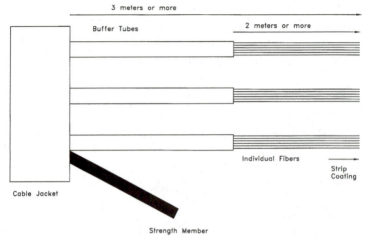

FIGURE 11.6   Cable stripping lengths.

buffered cable, ensure that the individual fibers with their 900-mm buffer are exposed and loose. Be careful not to damage the optical fibers.

4. Carefully clean all fibers in the tube of any filling gel that may be present in the cable with the proper gel remover. Use gloves to protect your hands from the cleaning solution.

5. Secure the end of the tube to the splice tray and lay out cleaned and separated fibers neatly on the table. Strip and clean the other cable tube's fiber that is to be spliced, and secure to the splice tray.

6. Identify the fiber to be spliced. Using the proper fiber coating stripper, remove enough coating so that about 5 cm of bare fiber cladding is exposed. For a tight-buffered cable, remove 5 cm of buffer first with a 900-mm buffer stripping tool, and then remove the 5 cm of coating with the coating stripping tool. This length is approximate and depends on cleaver requirements and splicing method. To help hold the small fiber securely in hand while stripping, clasp the fiber with a small piece of folded emery cloth. Always keep the coating stripping tool perpendicular to the fiber while stripping (see Fig. 11.7).

7. Carefully clean the bare fiber by wiping the fiber in one direction with a residue-free alcohol wipe (see Fig. 11.8). After cleaning, do not touch the bare fiber with your fingers, and prevent it from touching any surface.

8. Prepare the cleaving tool and adjust the cleaving length as required by the splicing technique (see the manufacturer's splicing instructions).

FIGURE 11.7   Optical fiber stripping technique.

FIGURE 11.8   Optical fiber cleaning technique.

9. Use the cleaving tool and cleave the fiber to obtain the perpendicular end surface. All fibers to be spliced must be cut with a cleaver. Using tweezers, immediately dispose of the cut fiber in a closable container designed for this purpose. (Caution: Wear safety glasses during cleaving procedure.)

10. Likewise, strip, clean, and cleave the other optical cable fiber to be spliced.

11. (a) For fusion splicing, place a splice protector, if used, onto one of the fibers to be spliced. Place both fibers into the fusion splicer and follow the fusion splicer's splicing instructions. A successful splice will be mechanically strong and will have a splice loss of less than 0.1 dB. Protect the splice with the splice protector (usually a heat-shrink or crimp-on type).

11. (b) For a mechanical splice, place and align the fibers in the mechanical connector. Follow the manufacturer's splicing procedure. A successful splice will result in a splice loss that is less than that specified by the manufacturer (typically less than 0.7 dB) and will be mechanically strong. To reduce possible fiber breakage due to torsion stress, secure the fiber tube in the splice tray if possible and then loop the fiber in the splice tray to create a permanent installation before splicing.

12. After the splice is complete, carefully place the splice into the splice tray and loop excess fiber around its guides. Ensure that the fiber's minimum bending radius is not compromised and keep all bends as smooth and as large as possible.

13. At this point, an OTDR test (or power meter test) of the splice can be performed while the splicing crew is still on site. Redo the splice if required.

14. After all fibers have been spliced, carefully close the splice tray and place it into the splice enclosure. At all times, ensure that the fibers' minimum bend radius is not compromised. Wrap excess fiber tube or buffered fiber around splice enclosure tube brackets, as shown in Fig. 11.1. Secure the fiber cable and strength member to the splice enclosure. Provide all grounding where required.

15. Test the splice with an OTDR (or power meter) from both directions. Refer to Sec. 13.4 on fiber optic facility measurement.

16. Close and mount the splice enclosure if all splices meet the engineering specifications.

11.5   OPTICAL FIBER TERMINATION

There are two different techniques used to terminate an optical fiber. Both are common throughout the industry. The field-installable connector technique is the process of terminating an optical fiber directly with a connector. Specially designed fiber optic connectors are installed directly onto the cable's fiber. The pigtail technique uses a factory-assembled optical fiber pigtail to terminate the fiber. Both techniques have their applications and advantages.

Field-Installable Connector. The field-installable connector technique allows for direct termination of optical fibers using special connectors designed for this purpose. The installation procedure involves securing the connector onto the cable's optical fiber with epoxy and then polishing the connector end to provide a low-loss connection. The end product is a cable with connectors installed directly onto each fiber.

The advantage of this technique is that a splice is not required for the termination. The cost of a splice and a splice tray is eliminated. The connectors are also less expensive than pigtails, but they take longer to install, thereby increasing labor cost. For an installation using a tight-buffered cable with a low fiber count, this technique can be attractive (see Fig. 11.9).

For multimode fiber, the resultant quality of the connection is usually good. The disadvantage of this technique, however, is that it is time-consuming; curing of the connector glue and meticulous polishing of the fiber end are required. This technique is also not popular for single-mode fiber termination. Because of the very small core size of single-mode fiber (10 μm), a good fiber-end polish may be difficult to achieve in the field. The quality of the connection depends considerably on the technique used by the installer. Factory-prepared pigtails are preferable and are commonly used instead.

Connectors can be installed on both tight-buffered and loose-tube fiber optic cable. However, extra care should be exercised when terminating loose-

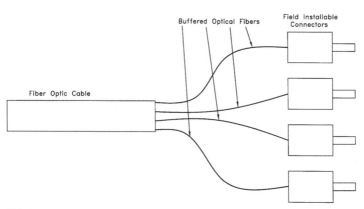

FIGURE 11.9   Field-installable connector.

tube cable with connectors. Loose-tube optical fibers are not protected by a buffer and can break easily if not handled properly. A fiber buffer sleeve can be slipped onto each loose-tube fiber to give it additional protection and support.

Field-Installable Connector Tool Kit

Ruler

Alcohol cleaning solution

Cable gel cleaning solution

Cotton swabs, no-residue

Paper tissues, no-residue

Fiber scribing tool

Fiber coating stripper

Fiber buffer and tube stripper

Fiber jacket stripper (specify jacket size)

Utility knife

Diagonal cutters

Scissors, serrated blade

Epoxy

Polishing film

Polishing jig

Polishing table

Heat gun (may not be required)

Crimping tool (may not be required)

Microscope

Proper fiber optic connectors

OTDR or power meter and light source

Large table and chairs

Field-Installable Connector Method. Manufacturers' installation techniques can vary for each connector type. The installer should become familiar with and aware of the specific manufacturer's procedures for installing the connectors. The following describes a general procedure for a field-installable connector:

1. Identify all optical fibers that will be terminated with connectors. Plan the exact fiber route to the equipment or patch panel. Ensure that the proper types of connectors are used.

2. Strip back about 2 m (6.6 ft) of the cable jacket to expose the fiber tubes or buffered fibers (the exact length will vary for different types of splice enclosures). Use cable rip cord to cut through the jacket. Then carefully peel back the jacket and expose the insides. If rip cord is not available, carefully strip the jacket with a jacket stripping tool or utility knife, making sure not to damage the optical fiber tubes or buffers. Cut off the excess jacket. Clean off all cable gel from exposed tubes or buffers with cable gel remover. Use gloves to protect hands from cleaning solution. Separate the tubes and buffers by carefully cutting away any yarn or sheath. Leave enough of the strength member to properly secure the cable in the splice enclosure (if required).

3. Slide the crimping collar onto the buffered fiber and/or heat shrinking tube as required.

4. Strip away about 4 cm (1.6 in) of fiber buffer as required by the manufacturer with a 900-mm buffer stripping tool.

5. Strip away about 2 cm of fiber coating (as required by the manufacturer) with a coating stripping tool.

6. Clean the exposed fiber with alcohol fiber cleaning solution.

7. Wipe the epoxy onto the exposed fiber as indicated by the manufacturer.

8. Slip the connector onto the fiber and push up against the buffer.

9. Slip the crimping ferrule onto the back of the connector and crimp on.

10. Place a small bead of epoxy on the front tip of the connector around the exposed fiber.

11. Allow the epoxy to cure as recommended by the manufacturer.

12. After the epoxy has cured, use the scribing tool and scribe (scratch) the fiber at the tip of the connector. (Caution: Wear safety glasses during fiber scribing.)

13. Break the scribed fiber end away from the connector by pulling the fiber straight away.

14. Screw the connector onto the polishing jig.

15. Using the proper polishing film, place the film onto a smooth, flat table. Then with light pressure, polish the connector using a figure 8 sweeping motion (see Fig. 11.10). This technique may need to be performed with two different grades of polishing film—fine and coarse. Check with the manufacturer.

16. After six or seven passes, examine the fiber end under the microscope. It should be free of scratches and glue. If scratches exist, continue polishing.

17. Test the connector for loss using an OTDR or a light source and power meter.

Pigtail Termination.   The pigtail termination technique involves the splicing of a factory-assembled pigtail onto a cable fiber. This ensures a quality connector installation with low optical power loss and low return loss at the connection. The factory-assembled connector provides the lowest possible optical loss, the best reliability, and the minimum return loss for both single-mode and multimode fiber. Because of the required splice, the splice loss should be factored into the link budget.

A splice tray and enclosure are used to house the splice and connector (See Fig. 11.11). The pigtail can be any length, allowing for optical fiber routing through equipment racks. This is the quickest way to terminate a fiber optic cable and can save significant time in large-fiber-count cable terminations. This method is popular for single-mode fiber or loose-tube cable terminations.

The drawbacks of this method are the higher cost of a pigtail compared to a field-installable connector, the need to make a splice in the fiber, and the requirement that a splice tray and patch panel or splice enclosure housing be used.

Pigtail Termination Tool Kit

Splicing tool kit (see Sec. 11.4)

Fiber optic pigtails

OTDR or power meter and light source

Splice tray

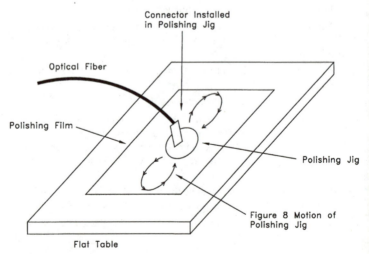

FIGURE 11.10   Field-installable connector polishing setup.

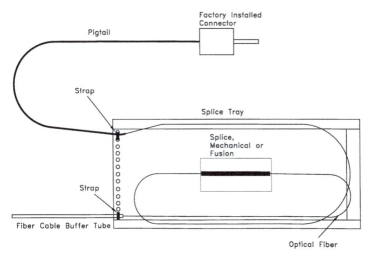

FIGURE 11.11   Pigtail termination.

Pigtail Termination Procedure

1. Identify all optical fibers to be terminated with pigtails. Plan the exact fiber route to the equipment. Determine fiber placement in splice tray, enclosure, or patch panel. Ensure that the proper connectors are used.

2. Prepare the cable end for splicing. Expose the individual fiber to be spliced. Clean, strip, and cleave the fiber (as described in Sec. 11.4, on splicing procedure).

3. Prepare the pigtail for splicing. Clean, strip, and cleave each fiber pigtail (as described in Sec. 11.4).

4. Splice the pigtail onto the cable fiber (see Sec. 11.4).

5. Mount splices in the splice tray and test with OTDR or power meter.

6. Install splice tray into splice enclosure or patch panel.

11.6 FIBER OPTIC CABLE TERMINATION

Fiber optic cable can be terminated in a number of different configurations.

Termination without Enclosure.   Fiber optic cable termination without an enclosure is the simplest and lowest-cost (material) type of termination. It is used mainly for terminating indoor tight-buffered cables with low fiber counts (normally less than 6). This type of cable is light and flexible and can be run directly to terminating lightwave equipment. Each cable fiber is terminated directly with a field-installable connector (see Fig. 11.12).

The fiber optic cable end is prepared by stripping back about 1 m (3 ft) of cable jacket and other protective layers to expose the individual buffered fibers. A sleeve, which provides additional support and protection, can be slipped onto each buffered fiber. The fibers are then terminated using the field-installable technique described in Sec. 11.5. Finally, a furcation protective boot is added to the cable end to provide the fibers with some strain relief.

This cable termination technique can be used for low-fiber-count loose-tube cables. However, proper fan-out termination kits should be used (available from most fiber suppliers). Loose-tube cable fibers are bare, with little support, and can be broken or damaged easily. The fan-out kit also includes fiber sleeves that can be slipped onto the individual fibers to provide them with protection and support. A furcation unit that provides proper loose-tube and cable termination is also included.

Whenever possible, the splice tray and an enclosure should be used for loose-tube termination, especially for heavy outdoor cables. Applications for this technique include dedicated LAN or video links with low fiber counts.

Termination in a Splice Enclosure. Termination in a splice enclosure allows for loose-tube or tight-buffered cable termination using the pigtail termination technique (see Fig. 11.13). It can be used with indoor or outdoor cables with higher fiber counts. The factory-assembled pigtails have protective jackets allowing pigtail routing through cabinets or racks so the cable can be connected to lightwave equipment. Splice enclosure termination provides an effective cable termination technique using fewer components than patch

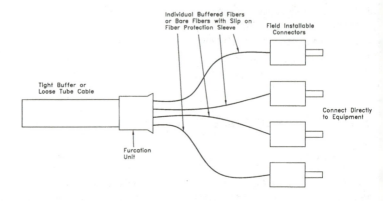

FIGURE 11.12   Termination without enclosure.

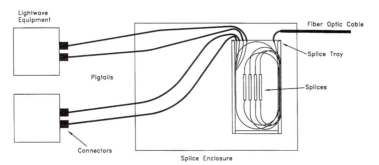

FIGURE 11.13   Splice enclosure termination.

panel termination (patch cords are not required) and eliminates a connection loss. However, it is not as versatile as patch panel termination.

Patch Panel Termination.   Patch panel cable termination is the most versatile and organized configuration. It provides quick and easy fiber identification and connection and allows patch cord connection or cross connection between equipment and other cable fibers.

The fiber optic cable can be terminated using the pigtail or field-installable connector techniques (see Fig. 11.14). The cable's optical fibers are spliced to pigtails that connect to the patch panel's bulkhead adapters. Fiber optic cable can be either loose-tube or tight-buffered with multimode or single-mode fibers.

For cable termination without splicing, the cable's optical fibers are terminated in the field with connectors and then connected to the patch panel's bulkhead adapters. This technique works best with tight-buffered multimode fiber using 900-μm or 3-mm buffered fiber. Precise field installation of connectors for single-mode fiber to achieve low connection loss may be difficult. Loose-tube cable can also be used in this configuration; however, the bare fibers should be protected with slip-on fiber sleeves and fanned out in a splice tray.

The following table provides a guide to the type of optical fiber jacket or buffer that is appropriate for various installation.

	Splice tray	Enclosure or patch panel	Cabinet, rack, or conduit	Indoors	Outdoors
Bare fiber	*				
Buffered fiber 900 μm	*	*			
Patch cord or 3-mm pigtail		*	*		
Fire-rated cable			*	*	
Outdoor cable					*

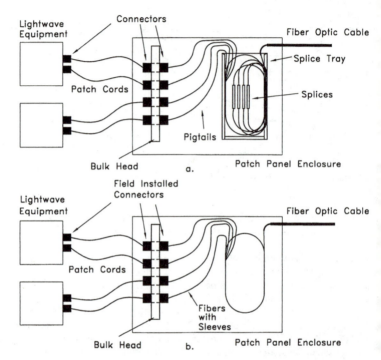

FIGURE 11.14 Patch panel termination.

CHAPTER 12
PATCH CORDS AND CONNECTORS

12.1 PATCH CORDS AND PIGTAILS

Fiber optic patch cords are analogous to electrical jumper cables. A fiber optic patch cord is a short-length optical fiber with a 3-mm tight buffer protective jacket and connectors at both ends (see Fig. 12.1). The jacket is colored orange for multimode optical fiber, and yellow for single-mode optical fiber. It is purchased factory-assembled in standard lengths or can be ordered in custom lengths. Patch cords have many uses, primarily to provide optical connections between installed lightwave equipment and fiber optic patch panels. It is also used for cross connecting fibers in patch panels or for connecting test equipment to fiber optic links. The patch cord's small diameter and flexibility allow for easy routing through tight locations in cabinets and racks full of equipment. The minimum bending radius for a patch cord is typically 3.8 cm (1.5 in).

Patch cords should be laid in cable trays that are dedicated to them and not left dangling where they may be inadvertently damaged. Tie wraps can be fastened snugly around the patch cords to secure them neatly. Excess patch cord lengths can be stored in appropriate storage trays or tied into smooth round loops with a radius larger than the patch cord's minimum bending radius. Since patch cord lengths cannot be adjusted in the field, it is best to measure the exact length required for the installation, and then order the custom-length patch cords from the supplier. This will eliminate an often messy situation where excessive lengths of patch cord slack are left dangling amongst equipment.

If a patch cord is cut in half, each half becomes a pigtail. A fiber optic pigtail is used to terminate an optical fiber with a connector. The pigtail is

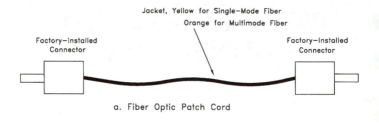

a. Fiber Optic Patch Cord

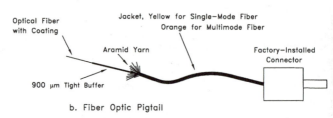

b. Fiber Optic Pigtail

FIGURE 12.1   Fiber optic patch cord and pigtail.

spliced (mechanical or fusion splice) to the optical fiber to provide a factory-quality connector termination.

Patch cords and pigtails should be selected to match the installed fiber optic cable's fiber type, and the type of connectors used at the patch panel and lightwave equipment. If the installed fiber optic cable uses a single-mode NDSF fiber with SC connectors, then the patch cord selected should be NDSF fiber with SC connectors.

12.2 LEGACY CONNECTORS

Over the years a handful of connector types have become common for most fiber optic installations. Because lightwave equipment manufacturers have not standardized on any one connector type, it is important to determine from the manufacturer the type of connector required. Often, manufacturers will provide a selection of connectors that can be fitted onto their equipment. This allows customers to maintain a connector standard throughout their fiber optic networks.

A connector is comprised of a ferrule, a body, a cap, and a strain-relief boot. The ferrule is the center portion of the connector that actually contains the optical fiber. It can be made from ceramic, steel, or plastic. For most connectors, a ceramic ferrule offers the lowest insertion loss and the best repeatability. The cap and body can either be steel or plastic. The cap can screw on, twist lock, or snap on to make a connection. The strain-relief boot relieves strains on the optical fiber.

The following list describes the various fiber optic connector types that have been established in the market for many years:

ST*   A good connector, popular for single-mode and multimode fiber connections, with an average loss of about 0.5 dB. It has a twist locking connection that is not susceptible to loosening in vibrating environments. It is a standard connector for most fiber optic LAN equipment. See Fig. 12.2*a*.

FC   A good connector, popular with single-mode fiber. Also known as FC-PC. It has a low loss with an average of about 0.4 dB or less. See Fig. 12.2*b*.

Biconic   This is an old-style connector. It was common for multimode and single-mode fiber systems but is now antiquated. It has poor repeatability and is susceptible to vibrations and high loss (over 1 dB). However, many installations still use these connectors. Replacement with higher-quality connectors such as FC or SC is a good idea. See Fig. 12.2*c*.

SMA   This is an older connector used for multimode fiber systems, but it still can be found on some equipment today. It has a higher

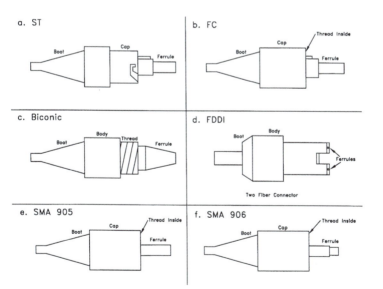

FIGURE 12.2 Common fiber optic connectors.

ST is a registered trademark of AT&T, USA.

loss of around 0.9 dB. Two types of SMA connectors are available, the SMA 905 and SMA 906. The SMA 905 has a straight ferrule (Fig. 12.2e). The SMA 906 has a stepped-down ferrule (Fig. 12.2f). The SMA 905 is available with a removable collar on the ferrule, needed to convert it into an SMA 906.

D4   This is a common connector type that is primarily used for single-mode fiber.

SC   This is a modular, higher-density connector. It has low loss (under 0.4 dB) and is common in single-mode installations.

FDDI   This connector is the FDDI standard fiber optic connector. It is a keyed duplex type, connecting two fibers at once. See Fig. 12.2d.

Bare fiber   This connector is used to connect an unterminated fiber. It is used to provide a temporary connection when testing bare fibers. It may require an index matching liquid to provide a low-loss connection.

A "PC" after the connector letter, as in FC-PC, means that the connectors make physical contact at the connection. This provides for a lower loss at the connection.

Many of these connectors are available with specially treated surfaces to minimize reflected optical light. There are three designation types: Super, Ultra, and Angle. The Super PC connector (written FC-SPC) has a curved surface designed to reduce reflected light. The Ultra PC connector provides further reduction in reflected light. If additional reduction in back reflections is necessary, the Angle connector (written FC-APC) can be deployed. It has the lowest reflectance specifications. This connector type is designated with a green connector boot. This connector type cannot be connected to a non-Angle connector because of its special angled surface. If this is done, the angled surface of the connector will be destroyed and the connector's special properties will be destroyed..

Super, Ultra, or Angle connectors are used in single-mode fiber applications with laser sources where reflected optical power can cause problems.

12.3 NEW SMALL FORM FACTOR (SFF) CONNECTORS

Small form factor (SFF) connectors are fiber optic connectors designed to maintain high connection specifications, such as low loss, and reduce the space occupied by connectors such as ST and FC. SFFs offer higher patch panel connector densities at a lower cost.

To date, five connector types have emerged in the marketplace (see Fig. 12.3). Externally, these connectors use a similar RJ-45 latch principle, which has been used expansively in copper applications. Small form factor connectors are as simple to use as the RJ-45s. As a result, the user is enabled to switch to fiber optic systems with ease. This also boosts the competitiveness of these systems in relation to copper.

Type	Developer
LC	Lucent Technologies
VF-45	3M
Opti-Jack	Panduit
MT-RJ	AMP, Siecor, Hewlett-Packard, US Conec, Fujikura
MU	Nippon Telegraph & Telephone (NTT)

LC   This connector's design resembles connectors based on zirconia-ceramic ferrules, which are used extensively today. Having been tested and used in patch panels, the LC has proven its suitability for multimode and single-mode applications. The connector is comprised of a latch mechanism similar to the RJ-45 latch and a ferrule with a 1.25-mm diameter. With its high packing density and proven connector technology, the LC connector is a forerunner for single-mode telecommunications applications (see Fig. 12.3*a*).

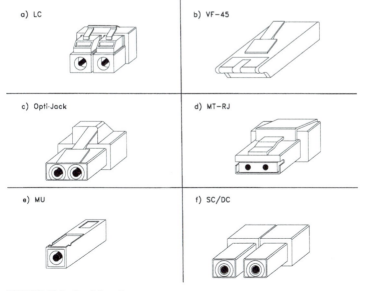

FIGURE 12.3   Small form factor connectors.

VF-45   The unique feature of this connector is that it has discarded
               the conventional ferrule technique. As a result, the VF-45 has
               the greatest cost-reduction potential of all the connector
               types. Another benefit for the replacement of copper systems
               by fiber using this connector is this connector duplicates the
               dimensions of the packing density and backplane section
               found in RJ-45 copper ports. The durability and reliability of
               this connector has been established through comprehensive
               testing and system installations (see Fig. 12.3b).

Opti-Jack   This is the industry's first RJ-45- style fiber optic connector
               (see Fig. 12.3c).

MT-RJ   The MT connector was formed initially as a multifiber con-
               nector used to couple transmit or receive diode arrays to mul-
               tifiber ribbon cables. To obtain the necessary accuracy of the
               fiber array, transfer-mold ferrule technology was developed.
               Dual-purpose, high-precision metal pins are inserted into the
               ferrule and used to provide an exact positioning mechanism.
               The MT-RJ connector is a revision of its larger MT sibling
               and represents a further development of the Mini-MT con-
               nector (see Fig. 12.3d).

MU   This connector system includes adapter-type optical connec-
               tors for cable connection (MU-A series), backplane connec-
               tors equipped with the self-retentive mechanism (MU-B
               series), simplified receptacles for connecting plugs to LD/PD
               modules (MU-SR series), and tools and accessories for oper-
               ation and maintenance.

               The MU-A connector consists of a 1.25-mm diameter zir
               conia ferrule and a push-pull coupling mechanism. This com-
               pact, high-density package is one-quarter the size of conven-
               tional fiberoptic connectors. The MU system has been
               designed for use in both plug-adapter-plug connector systems
               and backplane connector systems.

               Areas in which the MU system may expect to find appli-
               cation are: advanced optical transmission, exchange, and sub-
               scriber systems and other optical communications systems
               (see Fig. 12.3e).

SC-DC   This connector's uniqueness stems from the fact that its
               design has done away with the use of the latch of the RJ-type
               connectors. The SC-DC system has, however, made full use
               of the elements of the SC and SC-duplex latch.

               An extruded circular plastic ferrule with two or four fibers
               and a diameter of 2.5 mm is used, rather than the ceramic fer
               rule found in conventional SC connectors. The SC-DC also
               does not use metallic guide pins, giving it a cost advantage
               over the MT-RJ (see Fig. 12.3f).

12.4   *CLEANING CONNECTORS*

Small particles or dust can dramatically affect connector performance. A small flake of skin from the hand or scalp can easily be larger than the diameter of a single-mode core. It is therefore good practice to clean a connector each time it is to be connected (see Fig. 12.4).

Cleaning Items

Can of clean, compressed air with static-free nozzle

Clean, residue-free swabs (with small head in order to fit into adapter)

Lint-free cloth or tissue

Cleaning solvent, a residue-free alcohol solution (99 percent pure iso-propyl alcohol and distilled water)

Fiber Scope (100 to 200 power for multimode fiber and 400 power for single-mode fiber)

Method for Cleaning a Connector

1. Remove dust cap from connector and wipe with a clean, alcohol-dipped, lint-free tissue. Wipe the connector face and the ferrule.

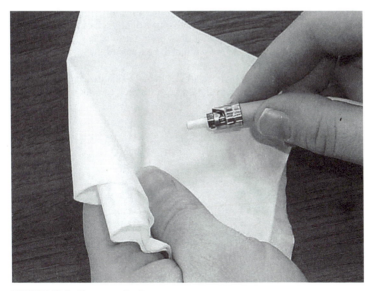

FIGURE 12.4   Connector cleaning.

2. Blow clean, compressed air over the entire connector surface.

3. Visually inspect the connector using a fiberscope for cleanliness. If not clean, repeat procedure.

4. Blow the dust cap with compressed air. Place the dust cap back on the connector.

Recently introduced by some vendors is a special connector cleaning tape that does a very good job in cleaning connectors. When using this cleaning tape, alcohol, cleaning tissue, and compressed air may not be required. Follow the manufacturer's instructions carefully for best results.

Method for Cleaning Adapters

1. Remove dust caps from both ends of the adapter, or remove connectors from the adapter. Insert a lint-free swab soaked with 99% alcohol solution into the adapter and clean the adapter.

2. Blow clean compressed air through the adapter.

3. Visually inspect the adapter for cleanliness. If not clean, repeat procedure.

4. Blow the dust cap with compressed air. Place the dust cap back on the adapter.

Only remove dust caps prior to making the connection. Do not let the connector touch any surface once the dust cap has been removed.

When making a connection, the connector should attach to the receptacle smoothly. Do not rotate the connector when making the connection. For screw-in connectors, tighten them only "finger tight." Never force a connector onto an adapter. If the connector does not want to mate, check to ensure the connector guides are properly aligned, and ensure that the proper connectors are being used.

When a connection is being made, extremely small diameter cores are being aligned, so the procedure should be performed carefully.

CHAPTER 13
POWER METER, SOURCE, AND RETURN LOSS MEASUREMENT

The fiber optic power meter and light source are used together to measure loss in a fiber or fiber optic device. The source launches the light into one end of the fiber, while the power meter is connected to the other end to measure the received optical power. The source can be an optical laser or light emitting diode (LED) designed as part of a test set, or alternately the lightwave communication equipment light source can be used.

Because optical fiber loss varies with light wavelength, power meter tests should be performed using the same wavelength as the one used by the lightwave communication equipment. If lightwave equipment operates at the 1310-nm wavelength, the power meter and light source should also be set to 1310-nm testing. If the lightwave equipment operates at 1550 nm, the power meter and source should also be set to 1550 nm. Special consideration should be made when testing through a WDM network; please see the section on WDM testing later in this chapter.

13.1   THE DECIBEL (dB)

The power meter test is used to determine light power loss in a fiber optic link. The measured unit of light power is the milliwatt (mW). However, a more convenient form of measurement used is called the decibel (dB).

The decibel is a common measurement used in the field of electronics to determine loss or gain in a system. It is the ratio, in logarithmic form, of power, voltage, or current levels between two points. One point is located at the beginning, or input, of the system to be measured, and the other point is located at the end, or output, of the system. The power formula for decibel gain is expressed as:

$$G_{(dB)} = 10 \times \log \text{ (output power/input power)}$$

When the output power is less than the input power, the value of this equation will always be negative. In most fiber optic applications, light power output from an optical fiber will always be less than the input light power into the optical fiber. Therefore, this value will always be negative. This negative gain can be referred to as a light loss, $L_{(dB)}$:

$$L_{(dB)} = -G_{(dB)}$$

where $L_{(dB)} = 10 \times \log$ (input power/output power).

Light loss, $L_{(dB)}$, is a commonly used specification for fiber optic attenuation. For example, to determine the light loss of an optical fiber in a cable, a light source is connected to one end of the fiber cable (input). The light output power of the source is known to be 0.1 mW. When an optical power meter is connected to the opposite end of the fiber optic cable under test (output), the meter measures 0.05 mW (see Fig. 13.1). Using the decibel power loss formula, the optical fiber loss can be calculated as follows:

$$L_{(dB)} = 10 \times \log \text{ (input power/output power)}$$

$$L_{(dB)} = 10 \times \log \text{ (0.1 mW/0.05 mW)}$$

$$= 3 \text{ dB}$$

The light power loss of this optical fiber is 3 dB.

The dB unit is a logarithmic ratio of input and output levels and is therefore not absolute (i.e., has no units). An absolute measure of power in decibels can be made in the dBm form. The dBm unit is a logarithmic ratio of the measured power to 1 mW of reference power.

The formula is as follows:

$$P_{(dBm)} = 10 \times \log \text{ (power/1 mW)}$$

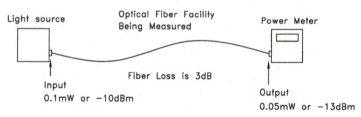

FIGURE 13.1   Optical power measurement.

The same result in loss can be achieved using the dBm. In the previous example, light power input by the source to the optical fiber is 0.1 mW, which is −10 dBm:

$$P_{\text{Source(dBm)}} = 10 \times \log (0.1 \text{ mW/1 mW})$$

$$= -10 \text{ dBm}$$

The light power received by the meter from the optical fiber's output is 0.05 mW, which is −13 dBm:

$$P_{\text{Receive(dBm)}} = 10 \times \log (0.05 \text{ mW/1 mW})$$

$$= -13 \text{ dBm}$$

The light power loss in the fiber is equal to the light source power minus the received meter light power:

$$L_{\text{(db)}} = P_{\text{Source(dBm)}} - P_{\text{Receive(dBm)}}$$

$$= -10 \text{ dBm} - (-13 \text{ dBm})$$

$$= 3 \text{ dB}$$

Therefore, the light power lost by the optical fiber is 3 dB.

All measurements must be in either decibels or in milliwatts, but not both. Usually, all measurements are made in the decibel scale because it is easier to work with. It is not necessary to convert between mW and dBm because most equipment specification data also use the decibel scale. The following table shows dBm equivalents for optical power in milliwatts:

Power in dBm	Power in mW
+20	100
+10	10
+3	2
0	1
−3	0.5
−10	0.1
−20	0.01
−30	0.001
−40	0.0001
−50	0.00001

It is helpful to remember that a loss of 3 dB is equivalent to a 50% loss in power. A loss of 10 dB is equivalent to a power loss of 90%. When you add or subtract a dB to or from a dBm, the result is in dBm. When you add or subtract two dB values, the result is always in dB. Decibel values are never multiplied together—they are always added or subtracted.

When measuring the loss in dB of a number of different sections of a fiber optic link, the total loss of all sections is equal to the sum in dB of each individual section.

Example 13.1.   A fiber optic link with a 1-km cable has a loss of 3.4 dB. Patch panel connection loss at each end is 0.8 dB. Pigtail loss is negligible. If a light source with optical power of −10 dBm is connected to one end of the fiber link, what will the received light power be at the other end? (See Fig. 13.2.)

First, the total link loss including patch panel connections is summed:

$$3.4 \text{ dB} + 0.8 \text{ dB} + 0.8 \text{ dB} = 5.0 \text{ dB}$$

The optical power loss formula needs to be rearranged to equal the received optical power:

$$L_{(dB)} = P_{Source(dBm)} - P_{Receive(dBm)}$$

becomes

$$P_{Receive(dBm)} = P_{Source(dBm)} - L_{(dB)}$$

$$= -10 \text{ dBm} - 5 \text{ dB}$$

$$= -15 \text{ dBm}$$

Therefore, the light power that would be measured by an optical power meter at the end is −15 dBm. It should be noted that two fiber optic connectors contribute to one connection loss.

13.2   TEST EQUIPMENT

For most power meter measurements, the following equipment is required:

Optical power meter:

• With proper wavelengths
• With proper connectors

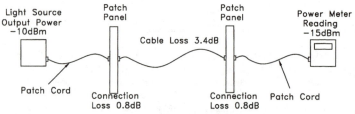

FIGURE 13.2   Fiber optic loss example.

- For single-mode or multimode fiber sizes
- Calibrated in dBm, preferably with optional dB scale
- Accuracy to at least ±0.1 dB

Optical light source:

- With stable light source
- With proper wavelengths (same as operating equipment)
- With proper connectors
- With single-mode or multimode fiber sizes
- With proper source, laser or LED source
- With sufficient output light power

Test patch cords:

- At least two, 1 to 5 m in length
- High quality with known loss (tested at factory and results documented)
- With proper connectors
- With same core size as fiber optic facility to be tested

Adapter:

- Good quality with ceramic sleeves
- Proper connector fitting
- Two required

Connector/adapter cleaning kit

13.3 PATCH CORD LOSSES

Before proceeding with optical power facility measurements, two high-quality test patch cords and two quality adapters (with ceramic sleeves) should be acquired. These patch cords should be accompanied by their factory-measured loss and reflectance values. The test patch cords and adapter should remain with the power meter and source at all times and be used only for testing. A quick check for test patch cord loss and connector alignment can be made as follows.

Method for Measuring Patch Cord Loss

1. Properly clean all connectors.
2. Turn on the power meter and light source test equipment and allow them to warm up and stabilize. If a laser light source is used, make sure that it remains off until all fibers to the source are properly connected.

3. Switch the power meter to the dBm scale. Connect test patch cord A, as shown in Fig. 13.3a, to obtain a reference light source power meter reading in dBm ($P_{\text{Ref(dBm)}}$). This value should be within light source output power specification. If it is not, the patch cord, power meter, or laser source may be faulty. Switch the power meter to the dB scale and set display to 0.0 dB (see the meter's instructions). Once display is set to zero, do not turn off or adjust the power meter.

   If the power meter does not have a dB scale and only displays the absolute power levels in dBm, record the power meter reading as $P_{\text{Ref(dBm)}}$. This is required for loss calculation in step 7.

4. Connect test patch cord B (as shown in Fig. 13.3b) between the light source and test patch cord A. A male-to-male adapter will be required for this connection.

5. Record the power meter light loss reading in dB ($L_{\text{Meter(dB)}}$). For some power meters, the light loss reading in dB may be a negative number. This indicates that the power meter uses the power formula instead of the light loss decibel formula for decibel gain (see Sec. 13.1). If the reading is negative dB, disregard the negative and use only the positive value for all calculations.

   If the power meter does not have the dB scale, record the dBm value as $P_{\text{Meter(dBm)}}$. Do not disregard the negative in dBm readings. A quick calculation is needed to determine the patch cord loss (see step 7).

6. Disconnect test patch cord B and flip it around, so that the connector now connects to the adapter, instead of the light source. Confirm that the power meter reading is the same as previously recorded. If the reading differs when it is reversed, try a different patch cord. Patch cord B's connectors may be out of alignment. Do the same for test patch cord A.

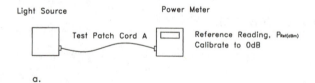

a.

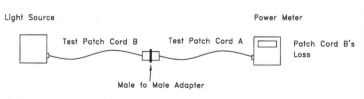

b.

FIGURE 13.3 Patch cord loss.

7. If the power meter reading is in dB, then this value can be used as patch cord B's loss ($L_{(dB)}$).

   If the reading is in dBm, subtract the received reading from the reference reading, as recorded in step 3, to obtain patch cord B's loss ($L_{(dB)}$):

$$L_{(dB)} = P_{Ref(dBm)} - P_{Meter(dBm)}$$

8. A good patch cord should have a loss of less than 0.7 dB and typically around 0.5 dB. The lower the better.

9. The patch cord loss is the optical loss of the patch cord's fiber and both connectors. Note that the loss of two connectors adds up to a connection loss.

13.4 FIBER OPTIC CABLE LOSS MEASUREMENT

A power meter loss measurement should be performed on all fibers in a fiber optic cable. The measured value is used to determine whether the cable's fiber loss is within equipment or design specifications. Two persons are required for this test, one at each end of the cable, as well as radio or telephone communications in order for them to coordinate the testing. The cable's fibers should be properly terminated with connectors for this test. The test measures the cable's fiber loss as well as loss of the connectors at both ends. In addition to this loss, the loss of the lightwave equipment fiber patch cords that span between the lightwave equipment and the patch panels (one at each end of the facility) should be added for complete link loss.

Fiber Optic Link Loss Measurement

1. Properly clean all connectors.

2. Turn on the power meter and light source test equipment and allow them to warm up and stabilize. If a laser light source is used, see Chap. 6 on safety precautions before beginning.

3. Identify the individual optical fibers to be tested. For an installed operating communication system, power down all connected communication equipment and ensure that all optical fiber light sources are off and completely disconnected from the fibers that will be tested.

4. Before the fiber facility can be tested for link loss, a reference reading should be recorded, and the power meter display should be set to 0 dB as follows:

   a). Using two test patch cords, A and B, connect the light source to the power meter as shown in Fig. 13.3b.

   b). Set the power meter to the dBm scale. Make sure that the light source is on, and read the received optical power at the power meter in dBm. This is the reference reading, $P_{Ref(dBm)}$.

c). Switch the power meter to the dB scale and set the display to show 0.0 dB.

d). Turn the laser source off and disconnect this test assembly but do not adjust or turn off the power meter.

e). If the power meter does not have a 0-dB setting and only displays the absolute power levels in dBm, then record the power meter reading $P_{\text{Ref(dBm)}}$ for later calculations.

5. Disconnect all existing equipment patch cords from the fiber optic cable patch panel's fibers that need to be tested. Connect the test patch cords, light source, and power meter as shown in Fig. 13.4 to the fiber facility to be tested. Do not disturb the power meter's zero reference calibration that has been adjusted in the previous step. Radio or telephone communication between the two facility ends will be necessary to coordinate the procedure events.

6. Ensure that the configuration is connected properly, and turn on the light source. Read the optical power meter and record the optical power level. Record the loss in dB as $L_{\text{Meter(dB)}}$. If the dB scale is not used or is not available, record the power level in dBm as $P_{\text{Meter(dBm)}}$.

7. The optical fiber link loss is the meter's reading in dB ($L_{\text{Meter(dB)}}$) and should be recorded. If the meter reading is only in dBm, the fiber optic total link loss is determined by subtracting the meter reading from the initial reference value $P_{\text{Ref(dBm)}}$:

$$L_{\text{(dB)}} = P_{\text{Ref(dBm)}} - P_{\text{Meter(dBm)}}$$

8. This procedure can be repeated to test all the fibers in the facility.

Example 13.2. An optical fiber is being measured for total link loss. The power meter is set to 0 dB using the light source and the test patch cords prior to the test. The power meter and source are connected as shown in Fig. 13.5 and the display shows 8.1 dB. Note that some power meters

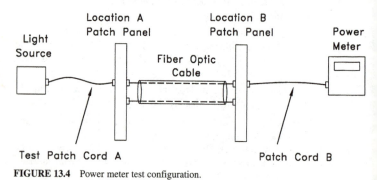

FIGURE 13.4 Power meter test configuration.

may show the dB loss value as a negative (-8.1 dB) in the dB scale. The meter derives this value from the gain formula (see Sec. 13.1). It should be converted to the positive loss value (8.1 dB) for use in all calculations. What is the fiber loss?

Because the power meter was set to 0 dB previously, the fiber link loss is 8.1 dB.

Example 13.3. The system in Fig. 13.6 was measured using a power meter that can only display values in dBm. Before the test, the two test patch Cords were connected together with the source and power meter and a reference value of -15 dBm was measured. The power meter and source are connected to the fiber cable facility as shown in Fig. 13.6 and a reading of -31.2 dBm is measured. What is the fiber loss?

$$L_{(dB)} = P_{Ref(dBm)} - P_{MeterEnd(dBm)}$$
$$= -15 \text{ dBm} - (-31.2 \text{ dBm})$$
$$= 16.2 \text{ dB}$$

Therefore, the total fiber link loss is 16.2 dB.

13.5 WDM LOSS MEASUREMENT

To do a proper test on fibers that connect to WDMs, the test set laser source wavelength must be set to the wavelength of the WDM's channel that needs to be tested, and the test set laser source spectral width must be within the WDM's channel spectral passband. WDMs have a narrow passband for each channel. The test equipment laser source spectral width must be within the WDM's channel passband, and at the proper channel wavelength, in order for the test to produce accurate results.

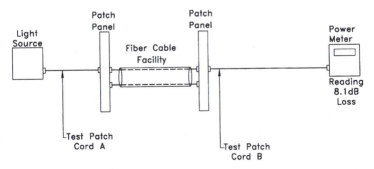

FIGURE 13.5 Power meter measurement example.

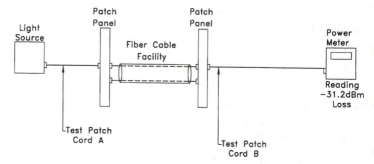

FIGURE 13.6   Power meter measurement example.

Often this type of test set laser source is not available. Therefore, the lightwave equipment designed to operate with the WDMs can also serve as the test laser source (for example, a SONET terminal laser). This will allow for an accurate fiber channel loss to be measured without purchasing an expensive laser test source.

WDM Channel Link Loss Measurement

1. Properly clean all connectors.

2. Turn on the power meter and allow it to warm up and stabilize. Identify the equipment laser source that will be used for the test. Review and observe all laser safety precautions (Chap. 6).

3. Identify the individual optical fibers to be tested and ensure that all optical fiber light sources are off and completely disconnected from the fibers that will be tested.

4. Before the fiber facility can be tested for link loss, a reference reading should be recorded and the power meter display should be set to 0 dB as follows

   a). Using two test patch cords A and B and adapter, connect the equipment light source to the power meter as shown in Fig. 13.7.

   b). Set the power meter to the dBm scale. Turn the laser source on and read the received optical power at the power meter in dBm. This is the lightwave equipment laser output power in dBm and is used as the reference $P_{\text{Ref(dBm)}}$.

   c). Set the power meter to the dB scale and set display to 0.0 dB (if available *c* see power meter manual).

   d). Disconnect this test assembly but do not adjust or turn off the power meter. If the power meter does not have a 0-dB calibration and only displays the absolute power levels in dBm, then record the power meter reading reference $P_{\text{Ref(dBm)}}$ for later calculations.

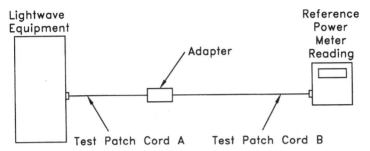

FIGURE 13.7   Reference level measurement.

5. Connect the test patch cords, equipment light source, and power meter as shown in Fig. 13.8 to the fiber facility to be tested. Do not disturb the power meter's zero reference calibration, adjusted in the previous step. Radio or telephone communication between the two facility ends will be necessary to coordinate the procedure events.

6. Read the optical power meter and record the optical power level in dB as $L_{Meter(dB)}$. If the dB scale is not used or not available, record the power level in dBm as $P_{Meter(dBm)}$.

7. The optical fiber link loss with WDMs is the meter's reading, $L_{Meter(dB)}$. If the meter reading is only in dBm, the fiber optic link/WDM loss is determined by subtracting the meter reading from the initial reference value $P_{Ref(dBm)}$:

$$L_{(dB)} = P_{Ref(dBm)} - P_{Meter(dBm)}$$

8. The resulting loss is the total end-to-end WDM channel loss, which includes fiber facility loss and WDM loss. To test a different WDM

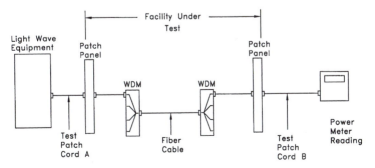

FIGURE 13.8   Fiber link/WDM loss using lightwave equipment laser source.

channel this procedure should be repeated using the lightwave equipment laser tuned to the wavelength for the channel to be tested.

Example 13.4. A channel of a DWDM SONET fiber system needs to be measured for total loss to see if it is within equipment specifications of 25.0 dB. The output of the SONET laser was measured to be 0.0 dBm. The SONET laser was then reconnected to the channel to be tested. Next, the power meter was connected at the SONET receiver end of the fiber channel and measured an optical level of -22.1 dBm. Is the WDM channel within optical budget?

$$L_{(dB)} = P_{Ref(dBm)} - P_{MeterEnd(dBm)}$$
$$= 0 \text{ dBm} - (-22.1 \text{ dBm})$$
$$= 22.1 \text{ dB}$$

Therefore, the total fiber link loss is 22.1 dB, which is within equipment specifications.

13.6 OPTICAL RETURN LOSS (ORL) MEASUREMENT

Optical return loss (ORL) is a measure of the total optical power reflected back to the light source end of a fiber. It is expressed in dB, as a ratio of reflected power over incident power.

$$\text{Optical return loss} = 10 \log (P_{Refl}/P_{Inc})$$

Optical reflections occur at various events in a fiber link, including at connectors, fiber ends, WDMs, and by the fiber material itself due to Rayleigh scattering. This optical power reflected back to the source end of a fiber can cause laser source instability and erratic operation (ORL measurements are not required for LED systems). ORL should be measured and compared to laser source specifications to ensure it is acceptable. If it is not acceptable, the various events that cause the high reflections should be identified and corrected or replaced.

ORL can be measured with an OTDR or ORL test set. An OTDR can provide accurate reflectance measurements of individual events along the fiber span. However, due to the OTDR's launch dead zone, reflectance from the connectors or events closest to the OTDR, which contribute the most to ORL, may not be seen or may be underestimated by the OTDR. Therefore, an ORL test set may be a better choice for this test.

An ORL test set uses the optical continuous wave reflectometer (OCWR) method to measure return loss. Continuous optical power is sent through a directional coupler to the fiber under test. The directional coupler routes the reflected light to a power meter. The reflected power is then measured and presented to the operator as a return loss.

To measure ORL, connect an ORL test set to the lightwave equipment transmit fiber using the installed equipment patch cords (using actual equipment patch cord and not test patch cord; see Fig. 13.9). Adjust the test set to the proper settings and record the ORL.

Fiber Optic ORL Measurement

1. Properly clean all connectors.

2. Turn the ORL source on and allow it to warm up. The ORL test set has a calibration procedure; follow the manual's procedure before performing this test.

3. Identify the cable's fiber that needs to be tested. Disconnect the transmit fiber patch cord at the equipment and not at the patch panel (review and observe laser precautions in Chap. 6).

4. If the fiber to be tested is connected to lightwave equipment at the far end of the cable, do not disconnect. The far end reflectance from the patch cord and equipment should be included in the measurement (note, the far end connection should only be to a receiver; disconnect if it is a light source).

5. Connect the ORL test set to the disconnected fiber patch cord, or if there is no patch cord, use a test patch cord, to connect the ORL meter to the fiber patch panel.

6. Read the optical return loss in dB.

If the reflected optical power is greater than transmission equipment specification, try switching to better-quality patch cords and connectors. Connectors with the Super or Ultra physical contact (SPC or UPC) designator have good return loss specifications. Angle physical contact (APC) connectors have better return loss specification but higher insertion loss.

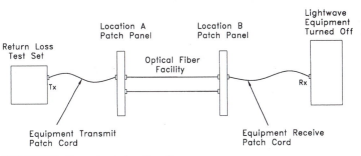

FIGURE 13.9   Return loss test configuration.

Remember that one only needs to test ORL at the end of the fiber that will be connected to a laser transmitter. ORL does not affect receivers. Contributions to ORL are greatest from events closest to the transmitter end of the fiber. This is because reflections from events far from the transmitter end of the fiber are attenuated by the fiber when propagating back to the transmitter end.

CHAPTER 14
THE OTDR AND OSA

14.1 THE OTDR

An optical time domain reflectometer (OTDR) is used to obtain a visual representation of an optical fiber's attenuation characteristics along its length. The OTDR plots this characteristic on its screen, in graph format, with the distance represented on the X-axis and attenuation represented on the Y-axis. Information such as fiber attenuation, fiber loss, splice loss, connector loss, reflectance, ORL, and anomaly location can be determined from this display. Note that the OTDR cannot determine fiber bandwidth limiting characteristics such as chromatic dispersion and PMD. It is strictly a tool for measuring and displaying fiber attenuation characteristics.

OTDR testing is the only method available for determining the location of a broken optical fiber in a fiber optic cable when the cable jacket is not visibly damaged. It provides the best method for determining loss due to individual splices, connectors, or other single-point anomalies installed in a system. It allows a technician to determine whether a splice is within specification or requires redoing. It also provides the best representation of overall fiber integrity.

In operation, the OTDR sends a short pulse of light down the fiber and measures the time required for the pulse's reflections to return to the OTDR and measures the reflection's optical power. Reflections along the fiber's length are caused by Rayleigh scattering and by single events such as connectors, fiber ends, splices, band bends, and other connected devices and fiber anomalies.

Knowing the optical fiber's index of refraction (n) and the time required for the reflections to return (T), the OTDR computes the distance to an event as follows:

$$\text{Distance}_{(meters)} = \frac{3 \times 10^8 \times T_{(seconds)}}{2 \times n}$$

The OTDR also measures the received optical power of the reflected light pulses and plots an optical fiber attenuation-by-distance display. This display, called a fiber trace, is a good representation of the fiber's attenuation characteristics along its length.

14.2 TEST EQUIPMENT

OTDR:

- Proper wavelengths
- Proper connectors
- Single-mode or multimode fiber sizes
- Sufficient dynamic range for length of fiber
- OTDR emulation software allows saved OTDR traces on diskette to be viewed on an office PC

  Test patch cords or pigtail:

- One or two, length as required
- Proper connectors
- Proper fiber size

  Connector cleaning solution, swabs, compressed air

  Bare fiber adapter

  Index-matching gel (for bare fiber adapter)

  Cleaver

  Cable and fiber strippers

  Dead-zone fiber (if required by OTDR)

Before proceeding with OTDR measurements, the OTDR should be checked to ensure that it has sufficient range capability to measure the entire optical fiber length. The following example shows a calculation that estimates the OTDR's fiber range in kilometers. The combined lengths of the fiber optic cable(s) to be measured should be less than this range.

Example 14.1

1. An OTDR manufacturer specifies that their equipment has a dynamic range of 25 dB [$D_{OTDR(dB)}$] at 1550 nm.
2. The optical fiber to be tested has attenuation of 0.25 dB/km at 1550 nm, which includes all splices.
3. Approximate OTDR range (dB):

$$\text{Range}_{(km)} = \frac{D_{OTDR\,(dB)}}{L_{(dB/km)}} = \frac{25\text{ dB}}{0.25\text{ dB/km}} = 100\text{ km}$$

This calculation is performed for all required operating wavelengths.

14.3 TYPICAL OTDR TEST METHOD

1. If the optical fiber to be tested is not connectorized, strip the fiber optic cable and expose a 2-m (6-ft) length of the fiber to be tested. Clean and then cleave the fiber to be tested.

2. Connect the OTDR to the fiber to be tested via a fiber patch cord or pigtail and bare fiber adapter. Also, add a dead zone fiber if required (see Fig. 14.1). A dead zone fiber is a small reel of optical fiber, up to 1 km in length (see OTDR specifications), inserted between the OTDR and the fiber under test. It is used with some OTDRs to allow the fiber under test to be moved out of the OTDR's launch dead zone, which can extend up to 1 km out of the OTDR. If a fiber event occurs in this dead zone, it would not be seen in the fiber trace. Some OTDRs do not require the use of a dead zone fiber; see OTDR instructions for details.

3. Ensure that a light source is not connected to the other end of the fiber that is to be tested.

4. Turn on the OTDR and let it warm up to a stable operating temperature.

5. Enter the proper OTDR parameters for operation, including wavelength, index of refraction of fiber to be tested, and pulse length (see OTDR operating instructions).

6. Start the OTDR test and allow the OTDR to average long enough to provide a smooth fiber trace.

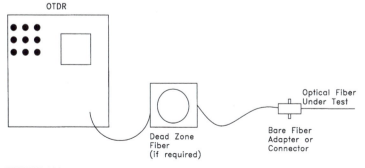

FIGURE 14.1 OTDR test configuration.

7. Adjust the resolution to display the complete fiber under test. To provide the best resolution, keep the pulse width as short as possible.

8. Measure loss for all anomalies, splices, connectors, and overall fiber.

9. Measure the fiber's total end-to-end loss in dB and fiber's attenuation in dB/km.

10. Store results and fiber traces on computer diskette or output results to a printer.

11. Repeat steps 1 through 10 for all required optical wavelengths.

Repeat the above steps with the OTDR connected to the other end of the fiber optic cable. Then average the two results for all events. This will provide a more accurate measurement of event loss and reveal any events near the beginning of the fiber that could not be seen because they were hidden by the OTDR's launch dead zone. This bidirectional testing method will also reveal events, such as splices, that are close to connectors, and hidden by the connector's Fresnel dead zone. Thus:

$$\text{Loss}_{\text{Event}} = \frac{\text{Loss}_{\text{EventDirectionA}} + \text{Loss}_{\text{EventDirectionB}}}{2}$$

14.4 READING OTDR EVENTS

An OTDR trace (Fig. 14.2) displays fiber loss and events along its length. The trace Y-axis is calibrated in dB and is used to read fiber and event loss. The X-axis is calibrated in miles or kilometers and is used to indicate the length of the fiber and fiber distance to an event. The sloped line plotted on the OTDR represents the fiber. The slope of this line is the fiber's attenuation.

At the beginning of a trace the line has a curved bump. This nonlinear region is the OTDR's launch dead zone and all events that occur in this zone will not be seen. In order to see events in this dead zone, the OTDR test should be conducted from the other end of the fiber, or a dead zone fiber can also be used.

In the linear section of the trace, accurate loss readings of fiber events can be read. A fiber splice is displayed as a small but sharp drop (Figs. 14.2 and 14.4a). The amount of the drop, as measured against the Y-axis in dB just before and after the drop, is the splice loss.

Occasionally, a splice event may be seen as a rise instead of a drop (Fig. 14.2). This is called a splice gainer and can be misinterpreted as light amplification, which it is not. A gainer is an OTDR splice event that occurs when two fibers from different stock or different manufacturers, and having different levels of Rayleigh backscatter, are spliced together. Light traveling into a fiber that has an increased level of backscatter will show up as a rise at the splice on the OTDR trace. If the OTDR test is performed from the other end of the fiber, then that splice will show as an exaggerated loss. Testing from both ends of the fiber and then averaging the result for the splice provides the more accurate loss.

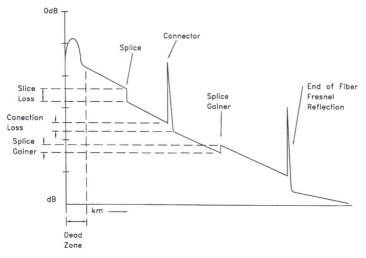

FIGURE 14.2 Fiber events.

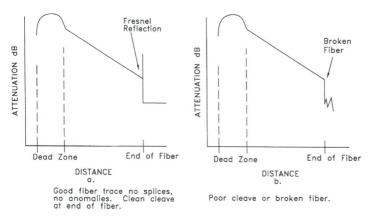

FIGURE 14.3 Typical OTDR traces.

$$\text{Loss}_{\text{Event}} = \frac{\text{Loss}_{\text{EventDirectionA}} + \text{Loss}_{\text{EventDirectionB}}}{2}$$

A 0-dB event can occur for the same reason as the gainer, except that the increase in backscatter exactly compensates for the splice loss. As a result, there will be no OTDR event and the splice cannot be located. Again, testing from the opposite end of the fiber will reveal the splice and its loss.

A fiber connector display is similar to a splice loss, except for a sharp upward spike just before the drop, caused by Fresnel reflection at the connector core-air boundary (Fig. 14.4b). Connection loss is the amount of the drop as measured against the Y-axis in dB just before the spike and after the drop.

Displayed at the end of the fiber trace is a sharp spike that is caused by Fresnel reflection due to the fiber-air boundary (Fig. 14.3a). The fiber length is the measurement from the beginning of the trace to the beginning of this Fresnel reflection spike. If there is no Fresnel reflection at the end of the fiber (Fig. 14.3b), either the end of the fiber has not been cleaved or the fiber has been broken.

14.5   DETERMINING EVENT PHYSICAL LOCATION

An OTDR can be used to determine an optical fiber's event location. However, the physical accuracy of the event in a cable is dependent on a number of factors: the accurate calibration of the OTDR, the pulse width setting, the accuracy of the manufacturer's supplied index of refraction (should be four digits or more) for the fiber core, and the accurate account for the amount of excess fiber in a cable.

The OTDR's pulse width is adjustable and can be set for larger values when testing longer fiber lengths. The larger pulse width has more energy

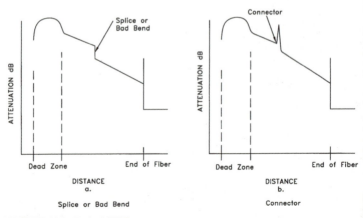

FIGURE 14.4   Typical OTDR traces.

and is therefore capable of probing longer fiber lengths. However, using a larger pulse width results in decreased event location accuracy. For best accuracy, the shortest possible pulse width should be used.

The fiber's core index of refraction, for the wavelength used by the OTDR, should be obtained from the manufacturer's fiber specifications and then entered into the OTDR. The OTDR uses this index of refraction value, along with the time it takes for the pulse to return, in order to calculate the distance along the trace. The index of refraction (n) should be accurate to at least four decimal places:

$$\text{Distance}_{(\text{meters})} = \frac{3 \times 10^8 \times \text{time}}{2 \times n}$$

There is an excess amount of fiber in a loose-tube cable due to the slight bunching of the fiber in the cable tubes and the tube's spiral wrap path around the cable central strength member. The cable manufacturer specifies this excess amount of fiber in the cable as a percentage of total cable jacket length. This excess fiber length should always be used to adjust the OTDR's fiber length.

Prior to the OTDR test, the fiber's index of refraction is entered into the OTDR, and the shortest pulse width is selected. Other OTDR parameters should be set as indicated by the OTDR operations manual. The OTDR is then connected to the fiber to be tested. An OTDR trace of the fiber is produced and the fiber distance to an event is measured. To account for the excessive fiber in the cable and to obtain an accurate cable jacket distance to an event, the OTDR's fiber distance to the event should be adjusted by the following formula (See Fig. 14.5.):

$$L_{\text{CableEvent}} = L_{\text{OTDREvent}}/(1 + a/100)$$

where $L_{\text{CableEvent}}$ = Cable distance to fiber event.

$L_{\text{OTDREvent}}$ = Fiber distance to event, as measured by OTDR

a = excess amount (in percent) of fiber in cable provided by cable manufacturer

Example 14.2. An OTDR trace shows that the distance to a fiber break in a cable is 55.72 km. The cable manufacturer specifies that the cable has 5.5% excess fiber. What is the cable jacket distance to the fiber break from the OTDR position?

$$L_{\text{CableEvent}} = L_{\text{OTDREvent}}/(1 + a/100)$$

$$= 55.72/(1 + 5.5/100)$$

$$= 52.82 \text{ km}$$

A comparative method can also be used if an accurate index of refraction or percent of excess fiber is not known.

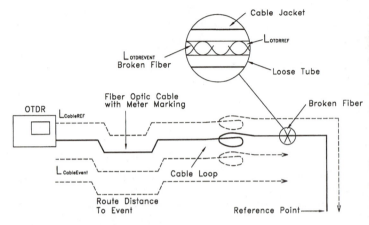

FIGURE 14.5   Fiber break location by OTDR.

Method

1. Using an OTDR, measure the distance to a known reference point in the cable, such as a splice or end of fiber location and record as L_{OTDRRef}.

2. Record the cable jacket length, using meter markings on the cable only, to this same reference point as in step 1, as L_{CableRef}.

3. Using the OTDR, measure the fiber distance to the event, and record as $L_{\text{OTDREvent}}$.

4. Calculate the cable distance to the fiber event as follows:

$$L_{\text{CableEvent}} = \frac{L_{\text{OTDREvent}} * L_{\text{CableRef}}}{L_{\text{OTDRRef}}}$$

The distance to the event is not the route distance, but only the fiber cable jacket distance (length) to the event.

14.6   THE OSA

The optical spectrum analyzer (OSA) is an instrument that can display the optical spectrum of a fiber. Its display is calibrated in dBm along its Y-axis and in nanometers along its X-axis. The instrument is used to display and measure characteristics of one or more optical signals of different wavelengths in the ᴍeasurements can include optical signal power, optical signal-to-noise ⁄NR), wavelength, spectral width, and signal channel spacing. It is a ⁄ tool when implementing or debugging WDM systems.

When connected to a fiber that carries multiple optical signals, the OSA display will be similar to Fig. 14.6. Each peak represents an optical signal. The height of the peak is the optical signal-to-noise ratio. The peak value, measured on the Y-axis in dBm, is that optical signal's power. The distance between the peaks is the optical signal's channel spacing.

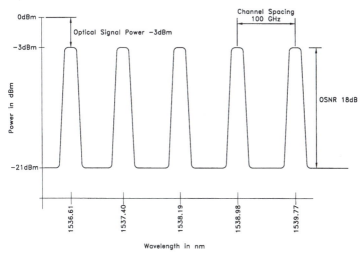

FIGURE 14.6   Optical spectrum analyzer display of a fiber with five signals.

CHAPTER 15
FIBER OPTIC INSTALLATION TESTS

15.1 FIBER OPTIC CABLE TESTS

A fiber optic cable should be tested three separate times during an installation:

1. The *reel test*. After the cable is received from the manufacturer and is still on the shipping reel, it is tested for manufacturer's defects or shipping damage. Any anomalies should be reported immediately to the manufacturer or shipper. The cable should not be installed until it passes this test.

2. The *splicing-installation test*. This test can be performed as soon as the cable is installed in the route, when all splices have been completed, and while the splice crews are still on site. It can identify any cable damage resulting from the installation process. It can also provide a splice-loss verification before splice enclosures are permanently mounted.

3. The *acceptance test*. This test is performed after the entire fiber optic system is complete and ready for commissioning. It provides the final commissioning data for engineering acceptance and archive data.

15.1.1 Reel Test

As soon as the cable reels are delivered, each fiber in the reel should be tested with an OTDR (see Fig. 15.1). This will establish that the optical fibers have been received in good condition from the manufacturer and are not damaged. OTDR fiber traces should be recorded and kept on diskette or in paper files.

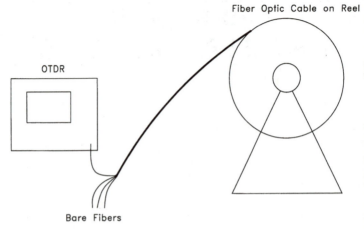

Fiber Optic Cable on Reel

OTDR

Bare Fibers

FIGURE 15.1   Fiber optic reel test.

Procedure

1. Loosen one free end of the fiber cable and strip it to expose all fibers (avoid unlagging the reel).

2. Strip and clean the individual fibers.

3. Using a bare fiber adapter, connect an OTDR to each fiber and record its trace. Use a deadzone fiber if required. For each fiber, record the following for all required optical wavelengths:

- Total loss
- Attenuation per kilometer
- Total fiber OTDR trace
- Any anomalies (zoom in on them and record them)
- Total reel length (marked on reel or obtained from cable markings)
- Total fiber length as indicated by OTDR
- Reel identification number, cable manufacturer, cable type, number of fibers in the cable
- Measurement direction
- Date
- Test equipment and serial numbers
- Crew members

All anomalies should be reported immediately. Each fiber should be free of anomalies and have no visible splice points.

4. After all fibers are tested, cut off the loose fibers and reseal the cable end to prevent entry of moisture and dirt. Secure the cable to the reel.

Confirm all results with the manufacturer's or engineering specifications. For installations where an OTDR is not available, an optical power meter test should be conducted to confirm fiber loss as specified by the manufacturer. Both fiber cable ends are required for this test.

15.1.2 Splicing-Installation Test

After each cable end is spliced, and while the splice crew is still on site, but before the splice enclosures are permanently mounted, the installed cable length and splices should be tested (see Fig. 15.2). If only one cable length is installed, this test can proceed immediately after termination of the fiber connection.

Tests are performed on each fiber, at all operating wavelengths, and from both directions.

Method

1. Two OTDR crews, in radio contact, prepare the far ends of both cables for testing.
2. The sequence of fibers to be tested is identified and testing is begun. A dead zone fiber is used if required.

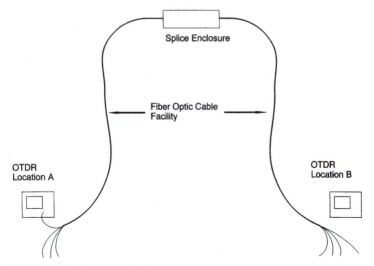

FIGURE 15.2 Fiber optic splice-loss test.

3. Each fiber is tested in one direction, then retested in the opposite direction. The following information is recorded:

- Total loss
- Attenuation per kilometer
- Total fiber trace
- Any anomalies (zoom in on them and investigate)
- Splice loss (gain) and splice trace
- Total cable length obtained from cable meter markings
- Total fiber length as indicated by OTDR
- Cable manufacturer, cable type, number of fibers in the cable, cable reel identification number
- Measurement direction
- Date
- Test equipment and serial numbers
- Crew members

Splice loss or gain and overall cable loss are tabulated for each fiber at each operating wavelength. Splice losses or gains from both OTDRs are recorded, and then the average is calculated to obtain an overall splice loss. If a splice gain is recorded, it is entered as a negative.

4. All bad splices should be identified and immediately redone. All fiber anomalies showing a loss greater than the engineering specification should be reported at once.

5. Newly installed cable lengths should have no optical fiber anomalies that have greater loss than specified. Ensure that there are no broken fibers.

6. After all fibers and splices have been tested to engineering specification, securely mount the splice enclosures and cable.

For an inexperienced crew, this test can be performed during the splicing procedure to confirm fusion splicer test results. The optical fibers should be retested once the splice enclosure has been permanently mounted.

15.1.3 Acceptance Test

Once installation is complete and the fiber is ready for equipment connection, a final acceptance test is conducted between connectors to ensure that the fiber optic link meets the engineering link budget specification. This test is usually conducted by the engineer or by technicians with engineers present.

The following test is done on the complete length of each optical fiber:

1. An OTDR is connected to one end of the fiber optic link.
2. The complete link is scanned, and the trace is stored. The following information is recorded at all operating wavelengths:

   - Total loss
   - Attenuation per kilometer
   - Total fiber trace
   - Any anomalies (zoom in on them and investigate)
   - Splice loss
   - Connector loss
   - Total link length obtained from cable meter markings
   - Total link length as indicated by OTDR
   - Cable manufacturer, cable type, and number of fibers in the cable
   - Measurement direction
   - Date
   - Test equipment and serial numbers
   - Crew members

3. An optical power meter and source are connected. Power meter readings are taken for each fiber at all operating wavelengths.
4. A return loss meter is connected to record the reflected power of fibers at both equipment terminating ends (if applicable).
5. A power meter measurement of equipment transmitter optical power is helpful for maintenance purposes. The output power of the lightwave equipment is recorded with a common output pattern, such as a digital all-ones pattern. Lightwave equipment power levels are measured at equipment output and receiver points. Data pattern, location, and levels are recorded.

15.2 FIBER ACCEPTANCE CRITERIA

Optical fiber acceptance criteria should meet engineering specifications.

Maximum Optical Fiber Attenuation in dB at a Wavelength:

$$\text{Loss}_{\text{max}} = d_{\text{cable}} \times L_{\text{dB/km}} + (L_{\text{dB/Splice}} \times N_{\text{splices}})$$

where d_{cable} = Length of cable in kilometers
$L_{\text{dB/km}}$ = Manufacturer's fiber loss specification per kilometer at wavelength

$L_{dB/Splice}$ = Maximum loss, per splice, at wavelength
$N_{splices}$ = Number of splices in cable segment being tested

Average Attenuation per Kilometer at a Wavelength:

$$\text{Average}_{dB/km} = \text{Loss}_{max}/d_{cable}$$

Maximum Splice Loss.   As per the engineering specifications, maximum splice loss depends on fiber type, splice type, and wavelength. Generally, for mechanical splices, the maximum splice loss is less than 0.3 dB. For fusion splices, it is generally less than 0.1 dB.

Maximum Connection Loss.   Connection loss depends on fiber type, connector type, and wavelength and is generally less than 0.5 dB. A good connection average loss would be about 0.3 dB or less.

15.3 *BIT ERROR RATE TEST*

A bit error rate test (BERT) is a point-to-point test performed to determine the quality of a digital system or channel (see Fig. 15.3). This test is generally conducted after all equipment has been completely installed and is in full operational mode, but before the system has been actually placed into service. A BERT can also be used on an existing communication system or channel, but the system or channel must be taken out of service to perform the test.

The test measures how many data bits in a transmission are received in error. The measure is a bit error rate (BER), which is the ratio of bits received in error to the total number of bits transmitted:

$$\text{BER} = \frac{\text{Number of received bits in error}}{\text{Total number of bits transmitted}}$$

A BER of 10^{-9} indicates that on average one error bit is received for every billion sent. A BER of 10^{-9} is adequate for many installations, including voice communications. For fiber optic systems, a BER of 10^{-12} is common.

The following procedure can be used to perform a BERT on a communication channel. The channel must be placed out of service for this test to be performed.

Method

1. The communications channel or system is set up and provisioned for testing.
2. Two BER testers are connected to each end of the channel using standard electrical interfaces such as RS232, RS449, V.35, DS1, DS3, and so on.
3. If only one BER tester is available, the opposite end of the link is physically looped back with a loop-back plug to complete the signal test path.

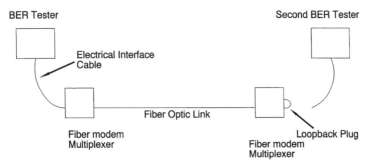

FIGURE 15.3   Bit error rate test (BERT).

If a loop-back plug is not available, then a software loop-back is used (however, the entire channel to the connector interface is not tested).

4. Both testers are configured and set to transmit a pseudorandom test pattern that will simulate live traffic, such as "2047 or $2^{23} - 1$."

5. The test is started and run for a specified period of time to determine the BER of the link. The testing period should be long enough to provide reasonable confidence of link performance. Link tests over one or two days are common for high-data-rate links. An event timer can be used to log erratic errors.

6. After the test period, the recorded BER should meet or exceed the system's BER specification.

15.4   RECEIVER THRESHOLD TEST

A receiver optical power threshold test is conducted to determine the lightwave receiver optical power threshold (see Fig. 15.4). This is the minimum light signal power required at the optical receiver to meet the equipment manufacturer's BER specifications for a digital system, or the signal-to-noise (S/N) specifications for an analog system. This value can be used when monitoring a receiver's performance and determining the link's optical margin.

The following procedure can be followed to measure the receiver optical power threshold.

Method

1. Determine the equipment's BER for a digital system or the S/N value for an analog system—as well as the receiver's optical threshold values—from the manufacturer's specifications.

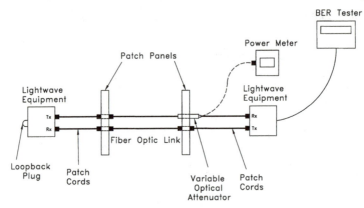

FIGURE 15.4 BER optical threshold test.

2. Take the lightwave equipment out of service for this test. For a digital system, connect a BER tester to the lightwave equipment and configure it for a standard BERT. For an analog system, set up an S/N test.

3. Conduct the BERT using a pseudorandom pattern, and record the system's BER. For an analog system, conduct the S/N test and measure the S/N.

4. Disconnect the patch cord at the lightwave receiver and connect it to an optical power meter. Measure the received optical power in dBm for the measured system BER or S/N value. Then reconnect to the receiver.

5. Insert a variable optical attenuator (VOA) between the optical receiver and transmitter.

6. Adjust the optical attenuator until the BER or S/N drops to the manufacturer's minimum specifications.

7. Disconnect the patch cord at the lightwave receiver and connect it to an optical power meter again. Measure the received optical power in dBm for the minimum BER or S/N value. This is the receiver optical power threshold, which should be compared to the manufacturer's specifications. Any further reduction in optical signal will degrade the system's performance below the manufacturer's specifications.

8. For a full duplex link, conduct this test in one direction and then repeat it again in the other direction to determine threshold values for both lightwave optical receivers.

CHAPTER 16
LIGHTWAVE EQUIPMENT

Lightwave equipment is a general term used here to refer to any optical fiber terminating equipment that converts electrical signals into fiber optic light signals, and fiber optic light signals back to their original electrical form. Lightwave equipment includes fiber optic converters, repeaters, or modems. The equipment can be very complex, requiring detailed installation instructions such as would be necessary with SONET/SDH terminals, or quite simple with few settings, such as those needed for an optical modem.

Lightwave equipment can also be combined with other communication equipment to provide a more integrated unit. In this case, this type of equipment is commonly referred to by its main function and has plug-in modules that provide the optical-to-electrical conversion. Two popular equipment types are multiplexers and LAN hubs. Both are multifunction units with available optical-to-electrical conversion modules.

Most optical fiber links use two fibers to provide full two-way communications. For some systems such as video links or public address, voice-only, one-way communication is required, so as a result, one fiber is used. The following table illustrates the various applications of links with one, two, and four fibers:

Application	One fiber	Two fibers	Four fibers
Voice (one-way, PA)	*		
Voice (telephone)		*	
Video (security)	*		
Video (interactive)	*		
Control systems (PLC)		*	
Telemetry	*	*	
Data communications		*	*
Multiplexer		*	*
Ethernet		*	
Token ring		*	
FDDI		*	*
SONET		*	*

WDMs can be used to reduce the number of fibers used for the application.

16.1 OPTICAL MODEM/MEDIA CONVERTER

The optical modem, also called a media converter, is a device that converts an electrical signal into an optical signal, and vice versa. Optical modems are available for most signals including Ethernet, Token Ring, RS232, video, and telephone. Figure 16.1 shows two examples of simple fiber optic communication systems using optical modems. Figure 16.1*a* is an extended video link, from camera to monitor, using one optical fiber. The system is quick and easy to install and normally does not require any adjustment with the modem. The video camera is connected to the transmit modem using the standard video coax cable. At the monitor end, the optical video signal is decoded by the receive modem and converted back to an electrical video signal. This system does not transmit audio, but there are modems that are available with audio connections.

In Figure 16.1*b* a simple modem system for computer communications is shown. The system is used to extend electrical communication such as RS232, RS449, and Ethernet. Two fibers are needed for bidirectional communication. Modems often allow the computer terminals to be moved further apart than standard electrical transmissions would permit (see modem specifications).

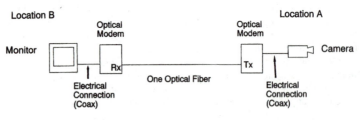

a. One-Way Transmission

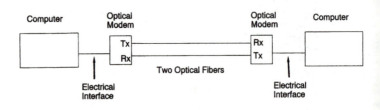

b. Two-Way Transmission

FIGURE 16.1 Optical modem link.

Figure 16.2 provides an example of a more complex modem link. Two computers, a telephone, and a video link use separate optical fiber modems connected through patch cords to a single fiber optic cable. Each modem is independent, meaning that the failure of one modem set will not affect other modems. Modem installation is usually quick and straightforward. Most just need to be powered on to be in full operational mode. Communication equipment troubleshooting is also simplified. One or both optical modems can be quickly replaced if suspected of failure. The system can be easily expanded by adding new modems to the fiber optic cable, assuming additional fibers are available. This type of configuration can be implemented for local inter- or intrabuilding communication. If many signals are to be transmitted over a longer distance, a more sophisticated approach (multiplexer system) that would reduce the optical fiber demand and increase the transmission bandwidth of a fiber should be considered.

The advantage of this configuration is that optical modems are relatively inexpensive, easy to use (many require little or no configuration), and available for many applications.

The disadvantage of the configuration is the need for dedicated fibers or fiber pairs for each signal. Optical fibers can be used up quickly in a cable.

16.2 MULTIPLEXER

A multiplexer is a unit that combines a number of electrical signals into one aggregate electrical signal. It can have built-in fiber optic transceivers or fiber

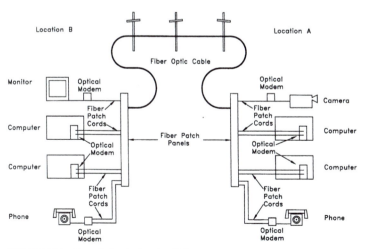

FIGURE 16.2 Expanded optical modem link.

optic transceiver cards that can be added to convert the multiplexer's aggregate electrical signal into an optical signal. Multiplexer systems make possible the more efficient use of both available optical fibers and bandwidth over fibers. A large number of communication signals can be placed onto two optical fibers. One such type of multiplexer is called a SONET terminal. SONET terminals that can multiplex more than 32,000 telephone conversations onto two fibers (for more on SONET/SDH, see Chap. 18) are available.

As shown in Fig. 16.3, multiplexers have the capacity to combine a number of different signal types such as voice, video, and data into one aggregate. Each communication signal is connected to a multiplexer electrical interface specific to the signal type. Video signals use coaxial cable. Point-to-point computer signals can use a number of interfaces, including RS232, RS449, or V.35. Telephone connections can be made with a standard two-wire POTS interface, E&M signaling, or a T1-DS1 connection.

Multiplexers can have network management capability. This allows an operator to monitor multiplexer performance and signal quality from a central location.

Multiplexers tend to be more expensive and complicated to install than optical modems, but the benefit of transmitting and receiving large amounts of signals over two fibers outweighs these deterrents.

16.3 OPTICAL AMPLIFIER

An optical amplifier is a unit used to boost optical signals without converting the signals into an electrical form. The optical signals remain as light signals,

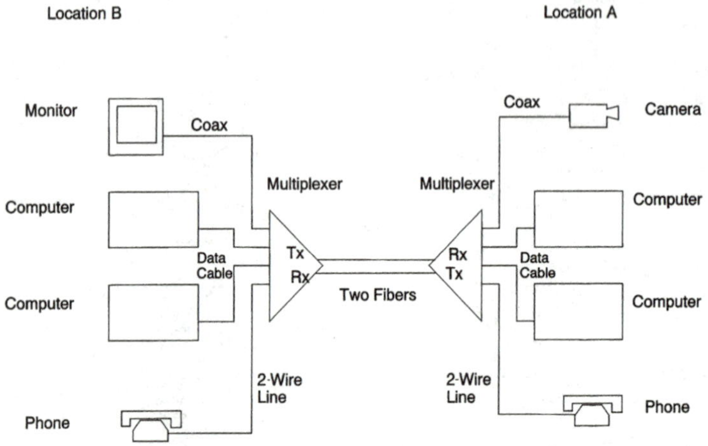

FIGURE 16.3 Multiplexer link.

at their source wavelength, throughout the amplifier. The amplifier can boost a band of optical wavelengths, not just one. Therefore, an optical amplifier is popular in WDM applications, where two or more wavelengths are present in a fiber.

An optical amplifier cannot, however, regenerate optical signals. Consequently, any distortion or other signal problems introduced in the fiber link are passed through the amplifier. Eventually the optical signals will need regeneration to clean them up. As a result, only a limited number of optical amplifiers can be used in any fiber optic link.

Optical amplifiers are available to boost optical signals in the 1310 and 1550-nm bands. The 1550-nm band amplifier is the most popular because of this band's lower fiber loss in long-haul transmissions.

The erbium-doped fiber amplifier (EDFA) is the most commonly used amplifier in fiber optic networks today. It can amplify an optical signal by as much as 35 dB in its operating band, 1530 to 1580 nm. It works on a principle of ion stimulation of erbium-doped fiber by a pump laser. A pump laser source in the amplifier unit, at 980 or 1480 nm, provides constant optical light that excites the erbium ions in the doped fiber to a higher energy state. The optical signal to be amplified is introduced into the end of the doped fiber.

As the optical signal photons pass through the doped fiber, they collide with the excited erbium ions, which causes the ions to return to their relaxed state. As a result, the erbium ion releases high-energy photons with the same wavelength and signal characteristics as the colliding optical signal photon. Therefore, the incoming optical signal is amplified.

Two common optical amplifier designs, the single-pump and dual-pump designs, are available today. The single-pump amplifier has a one-pump laser at the signal input of the erbium-doped fiber. The dual-pump version uses two pump lasers at both ends of the erbium-doped fiber. Typical amplifier gains are 17 dB for the single-pump and 35 dB for the dual-pump amplifier (see Fig. 16.4).

Optical amplifiers are available in three configurations—post-amp, line amp, and preamp. See Fig 16.5.

Postamp configuration requires the amplifier to be located at the optical signal source site. It is used as a power booster for the optical signal before the signal is launched into the outside plant fiber.

A line amp is placed at a midspan location to enable the amplifier to boost a weak signal, thus providing enough signal amplification for the signal to continue through the rest of the outside plant fiber.

The preamp is placed at the receiver end of the fiber link and is used to boost weak signals just prior to the optical receiver.

Optical output from an optical amplifier is intensely powerful, with the potential to harm the eye; extreme care should be exercised when working with this equipment to prevent eye damage. All equipment should be turned off before disconnecting optical fibers. Electrical power to the amplifier should be turned on only after all fiber connections have been properly com-

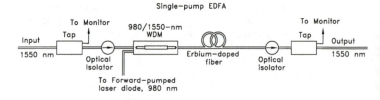

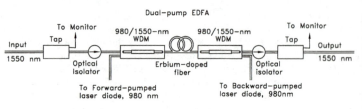

FIGURE 16.4 EDFA Optical amplifier block diagram.

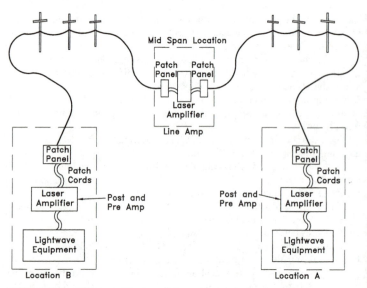

FIGURE 16.5 Optical amplifier in post, line, and preamplifier positions.

pleted. Powerful optical amplifiers should have a safety interlock that turns off the laser optical output if the fiber link is broken or disconnected.

When testing a fiber optic link, with OTDR or a power meter that has an optical amplifier installed, the optical amplifier should be turned off and disconnected from the fiber optic link. Each segment of the fiber optic link should be tested separately. Since an optical amplifier will permit optical signals to pass in only one direction, an OTDR requires bidirectional signal flow in a fiber (primary and reflected light pulse), and therefore will not work.

The following are typical optical amplifier specifications that should be considered for all amplifier link designs:

Specification	Units	Comments
Operating bandwidth	nm	Wavelengths that can be amplified
Gain	dB	Small signal gain
PMD	ps/$\sqrt{\text{km}}$	Polarization mode dispersion
PDL	dB	Polarization mode loss
Noise figure	dB	Due to spontaneous amplifier noise (ASE)

16.4 OPTICAL REGENERATOR

An optical regenerator boosts and regenerates an optical signal at a midspan location. The regenerator receives the weak optical signal and converts it to an electrical signal. The electrical signal is regenerated by removing effects due to fiber dispersion and other distortions, and then it is retimed, amplified, and finally converted back to an optical signal. Regenerators are often used to extend transmission distances over hundreds of miles.

16.5 LIGHT SOURCES

The two types of light sources used by lightwave equipment for optical fiber transmission are light-emitting diodes (LEDs) and lasers.

LEDs are economical and used mostly for short-distance low-data-rate applications (up to 125 Mbps), such as LAN communication. They are available for all three wavelengths but are most common at 850-1310 nm (850-nm LEDs are usually the least expensive). Light power from an LED covers a broad spectrum, from 20 to over 80 nm (see Fig. 16.6). The LED is more stable and reliable than a laser in most environments.

Lasers are more expensive than LEDS, but are more advantageous to use in the high-modulation bandwidth (over 2 GHz), with high optical output

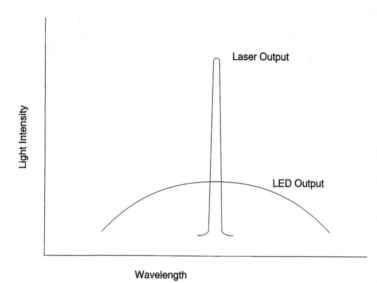

FIGURE 16.6   Light source spectrum.

power and narrow spectral width. The application of lasers is in long-distance, high-data-rate requirements. Lasers are common in single-mode optical fiber applications, and their light power covers a very narrow spectrum, usually less than 3 nm. This results in a low chromatic dispersion value, and hence high fiber bandwidth.

The life span of a laser is shorter than that of an LED. Lasers are sensitive to the environment (especially to temperature variation) and to such fiber design parameters as reflected optical power. Lasers are also more complex than LEDs and require greater overhead in terms of electronic and temperature control to maintain stable operation. Therefore, lasers are not often used in fiber to desktop or LAN applications.

The exception to this rule is the vertical cavity-surface-emitting laser (VCSEL). The VCSEL's manufacturing is similar to an LED but it operates as a laser. It also requires less supporting circuitry and therefore has a much lower cost than a standard laser. A VCSEL is presently available for the 850-nm band but may be released for 1310 nm in the future.

Optical output from a laser is strong and can cause injury to the eye. Never look into laser light or a fiber coupled to a laser. Ensure that all laser sources are powered off before disconnecting the fibers. Care should be exercised when working with laser sources. Note that fiber optic laser sources' light is not visible to the human eye.

LED and Laser Chart Comparison

	Wavelength	Spectral width	Output power	Data rate	Detector	Fiber type	Cost
LED	850/1310 nm	>50 nm	−15 dBm	<125Mbps	PIN	Multimode	Low
VCSEL	850 nm	<3 nm	0 dBm	<2 Gbps	PIN/ APD	Multi/ single mode	Moderate
Laser (FP)	1310/1550 nm	<3 nm	0 dBm	>2 Gbps	PIN/ APD	Single mode	High
Laser (DFB)	1550 nm	0.0004 nm (50 MHz)	0 dBm	>2 Gbps	PIN/ APD	Single mode	Very high

Chart uses typical values.

The spectral width is measured by the full-width half-maximum (FWHM) method. The FWHM is the width of the laser or LED spectrum between the wavelengths at which the light intensity is half that of the peak value.

Lasers used in long-haul transmissions, such as for SONET systems, are either Fabry-Perot (FP) or distributed feedback (DFB) type. Both have high-data-rate capacity and powerful optical signals.

The Fabry-Perot laser is more popular because of its lower cost. The DFB laser is used for its very narrow spectral width (which reduces chromatic dispersion) and its low noise characteristics. The DFB laser's narrow spectral width is necessary in DWDM fiber systems.

The disadvantage to the DFB laser is its very high cost and extreme sensitivity to back reflections. If DFBs are used, back reflections for the fiber link should be carefully managed.

16.6 OPTICAL DETECTION

Optical detection occurs at the lightwave receiver's circuitry. The photodetector is the device that receives the optical fiber signal and converts it back into an electrical signal. The most common types of photodetectors are the positive intrinsic negative (PIN) photodiode and the avalanche photodiode (APD).

PIN photodiodes are inexpensive but require a higher optical signal power to generate an electric signal. They are more common in short-distance communication applications.

The APDs are more sensitive to lower optical signal levels and can be used in longer-distance transmissions. They are more expensive than the PIN photodiodes and are sensitive to temperature variations.

Both photodiodes can operate at similar, high signal data rates. Some receiver photodetector circuits operate within a narrow optical dynamic range. The receiver's optical dynamic range is the light-level window in dBm through which a receiver can accept optical power. The receiver will only accept light within the manufacturer's predetermined levels (measured in dBm). If the light is too strong, the receiver's circuit will saturate, and the equipment will not operate. This is usually caused by insufficient fiber link attenuation. To remedy this problem, optical fiber attenuators are inserted into the link until the received optical power level is within the receiver's dynamic range. Attenuators can be installed at the equipment connector or at the patch panel connector on the receive or transmit fiber. For laser sources, ensure that the return loss measurement is within the equipment's specifications.

If the light is too weak, the receiver will not be able to detect the signal, and the equipment will not operate. In this case, attenuation must be removed from the fiber optic link. Many receivers designed to operate for short-distance communication, such as for LAN communication, are designed not to saturate when full light power level is received (no link attenuation).

CHAPTER 17

WDMs AND OTHER
OPTICAL COMPONENTS

A wavelength division multiplexer (WDM) is a unit that couples different optical wavelengths (λ) from two or more fibers into one common optical fiber. A WDM can also recover the different optical wavelengths, in the common fiber, into separate fibers. It can be a completely passive unit with the ability to fit into a small cassette, or it can be integrated with electronics that provide additional functionality for the unit such as remote power monitoring or amplification. WDMs are used in pairs, with one WDM at each end of the fiber. At the one end, a WDM couples different optical wavelengths into a fiber, and at the other end a compatible WDM recovers the different wavelengths and directs them into separate fibers (see Fig. 17.1).

Each channel in a WDM is designed to pass a specific optical wavelength or wavelength band. For example, a two-channel WDM system can be specified for the 1310- and 1550-nm bands. This system can be used to place two optical signals into one common fiber. An optical signal from a 1310-nm wavelength source is connected to the WDM's 1310-nm channel, and the optical signal from a 1550-nm wavelength source is connected to the WDM's 1550-nm channel. At the opposite end of the fiber, a compatible WDM is used to recover the two wavelengths into two separate fibers. A four-channel WDM system can be specified for four wavelengths (see Fig. 17.1). A WDM will couple four wavelengths into one fiber, and a compatible WDM will recover the four wavelengths into separate fibers at the opposite end.

The WDM channels act like filters that pass only optical signals for each specified wavelength band, so an attempt to connect a 1310-nm optical signal to a 1550-nm channel will not work. Filters most commonly used in WDMs are fiber Bragg gratings (FBGs). A fiber Bragg grating is a short piece of fiber that has a reflection filter permanently written into its core. The filter is created by gratings formed by a periodic modulation, using ultraviolet radiation, of the refractive index of the core during the fiber manufacturing processes. An

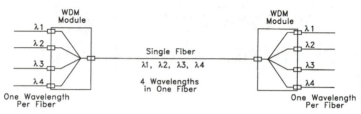

FIGURE 17.1   Simple WDM configuration.

FBG reflects wavelengths that the FBG is designed to reflect, and passes all other wavelengths. Fiber Bragg gratings have three uses in fiber systems. FBGs are used in WDMs, allowing multiple wavelengths of light to be combined on to a single fiber. With FBGs, specific wavelengths from a fiber can be added or removed; therefore they are used in add or drop type WDMs. Also, with FBGs, the light source can be controlled and reach a wavelength precision accuracy of ±0.02 nm.

At the present time, WDMs are most commonly available either with 2, 4, 8, 16, 32, and 64 channels. WDMs with a capacity of greater than 64 channels are in the process of being developed.

A two-channel WDM that couples wavelengths from the 1310-nm and 1550-nm bands is referred to as a wide-band or cross-band WDM. A narrow-band WDM is a WDM that couples two or four wavelengths onto a fiber only in the 1550-nm band. Therefore, transmitters used with a narrow-band WDM are all tuned in to wavelengths in the 1550-nm band. A dense WDM (DWDM) is a type of narrow-band WDM that is designed with 100-GHz (0.8-nm) channel spacing, and due to this closer channel spacing, it can couple eight or more wavelengths into the 1550-nm band (see Fig. 17.2).

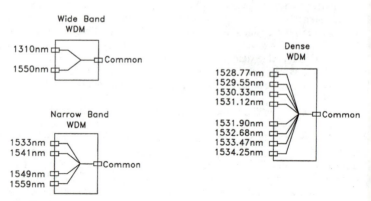

FIGURE 17.2   Wide-band WDM, narrow-band WDM, and DWDM.

It is imperative that WDMs be used with different wavelength laser sources and be tuned to the specific wavelength or band specified for WDM. For our example, when using 1557- and 1533-nm narrow-band WDMs, 1557- and 1533- wavelength laser sources must be used (see Fig. 17.3). If proper matching wavelength laser sources are not used, the system may not function properly.

Prior to the introduction and implementation of WDMs, two fibers were required in a communication system (see Fig. 17.4). One fiber would be connected to the communication unit optical transmitter while the other fiber was connected to the optical receiver. This enabled the flow of information in two directions at the same time between communication units. This is known as full duplex communication.

Now, using WDMs, only one fiber is needed for the facilitation of full duplex communication. However, the optical transmitters of the communication system need to be tuned for the wavelengths specified for the WDMs (see Fig. 17.5).

If a four-channel WDM is used, then two communication systems can be placed into one fiber. If an eight-channel WDM is used, then four

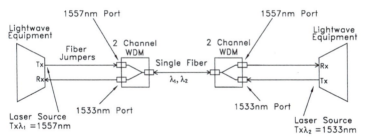

FIGURE 17.3   1557- and 1533-nm WDM system.

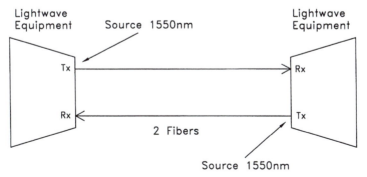

FIGURE 17.4   Two-fiber communication system.

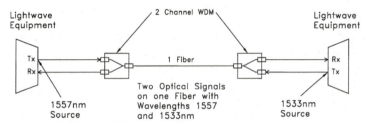

FIGURE 17.5   One-fiber communication system with WDMs.

communication systems can be placed into one fiber (see Fig. 17.6). It is clear to see that the use of WDMs helps to greatly reduce the number of fibers required for communications systems, and hence reduce the costs of a fiber cable plant.

WDMs are specified by their channel wavelengths in nanometers (nm) and their transmit-and-receive configuration. Occasionally the channel wavelength may be specified as a frequency in terahertz (THz). The relationship between frequency and wavelength is as follows:

$$F\ (\text{THz}) = \frac{299792}{\lambda\ (\text{nm})} \quad \text{or} \quad \lambda\ (\text{nm}) = \frac{299792}{F\ (\text{THz})}$$

The value 299792 is the estimated value for the speed of light in glass, with its decimal place adjusted for the proper units for this formula. The speed of light in glass is approximately 2.99×10^8 m/s.

A WDM that has two or four channels in the 1550-nm band is referred to as a narrow-band WDM. It has channel spacing of 1000 GHz ($\sim$8 nm).

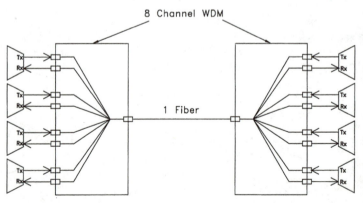

FIGURE 17.6   One fiber used for transmission for four systems.

WOMs with 8, 16, 32, or more channels are referred to as dense WDMs (DWDMs) because in order to achieve this number of channels, the channel spacing is reduced to 200 GHz or less.

The industry has standardized on DWDM channel wavelengths and spacing between wavelengths as recommended by the International Telecommunications Union (ITU). The ITU-T standard defines DWDM channels with 100-GHz (~0.8-nm) channel spacing referenced to the krypton line at 193.1 THz.

The following charts show the industry standard 1000-GHz spacing frequencies used in narrow band WDM applications, and 100 GHz used in DWDM applications.

Narrow-Band WDM 1-THz (1000-GHz) Channel Spacing

Wavelength, nm	Frequency, THz	Wavelength, nm	Frequency, THz
1533.47	195.5	1549.32	193.5
1541.35	194.5	1557.36	192.5

DWDM 100-GHz Channel Spacing

	Channel	Wavelength, nm	Frequency, THz		Channel	Wavelength, nm	Frequency, THz
	65	1525.66	196.5		40	1545.32	194.0
	64	1526.44	196.4		39	1546.12	193.9
	63	1527.21	196.3		38	1546.92	193.8
B	62	1527.99	196.2	R	37	1547.72	193.7
L	61	1528.77	196.1	E	36	1548.51	193.6
U	60	1529.55	196.0	D	35	1549.32	193.5
	59	1530.33	195.9		34	1550.12	193.4
E	58	1531.12	195.8	I	33	1550.92	193.3
I	57	1531.90	195.7	L	32	1551.72	193.2
S	56	1532.68	195.6	O	31	1552.52	193.1
H	55	1533.47	195.5		30	1553.33	193.0
	54	1534.25	195.4	N	29	1554.13	192.9
O	53	1535.04	195.3	G	28	1554.94	192.8
R	52	1535.82	195.2		27	1555.75	192.7
T	51	1536.61	195.1		26	1556.55	192.6
	50	1537.40	195.0	B	25	1557.36	192.5
	49	1538.19	194.9	A	24	1558.17	192.4
B	48	1538.98	194.8	N	23	1558.98	192.3
A	47	1539.77	194.7		22	1559.79	192.2
N	46	1540.56	194.6	D	21	1560.61	192.1
D	45	1541.35	194.5		20	1561.42	192.0
	44	1542.14	194.4		19	1562.23	191.9
	43	1542.94	194.3		18	1563.05	191.8
	42	1543.73	194.2		17	1563.86	191.7
	41	1544.53	194.1		16	1564.68	191.6

The block of 1550-nm band wavelengths currently used for optical communications ranging from 1530 to 1565 nm is referred to as the "C" band. The C band is further subdivided into blue and red bands. The blue band includes the wavelengths from 1527.5 to 1542.5 nm, and red band from 1547.5 to 1561.0 nm.

The block of wavelengths from 1570 to 1610 nm, referred to as "L" band, is being investigated for future use. The block of wavelengths from 1525 to 1538 nm is referred to as "S" band and is also being investigated for future use.

Note that the ITU-T channel spacing is defined as frequency in terahertz. Therefore, to obtain the appropriate wavelength, the speed of light in the fiber must be divided by the frequency. Due to slightly different fiber glass composition in different fibers, the speed of light may differ slightly for different fibers. Hence the wavelengths shown here are only approximate and may vary slightly for different fibers.

The closer the channels are spaced together, the higher the number of channels which can be inserted into this band. Presently available technology now allows for 50-GHz channel spacing ability (see Appendix F). The channel spacing used in WDMs is important to note in network design. The closer together the channels, the smaller the spectral bandwidth for each channel, so the need for the laser spectral width to be tighter increases (see Fig. 17.7). Also, the laser wavelength must be stable enough during its lifetime so it does not drift outside the WDM's channel spectral bandwidth.

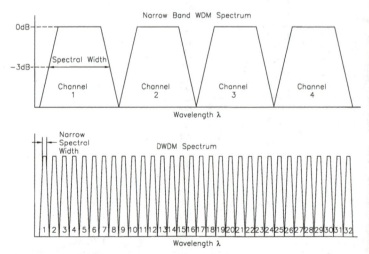

FIGURE 17.7   Example of a WDM channel specification.

A WDM that is configured to connect to only optical transmitters or only optical receivers is called a unidirectional WDM. A unidirectional WDM will only permit optical transmission in one direction in a fiber. The unidirectional WDM that connects to only optical transmitters is referred to as the mux, and the unidirectional WDM that connects to the receivers is known as the demux (see Fig. 17.8). A WDM that is configured to connect to both optical transmitters and receivers is called a bidirectional WDM. It is designed for optical transmission in both directions in one fiber (see Fig. 17.9).

Bidirectional WDMs that allow optical transmission in both directions through the same WDM channel (port) are also available (see Fig. 17.10). They are referred to as universal and are identical and interchangeable. The greatest benefit of the universal WDMs is their versatility; they can be inserted at either end of the fiber link and require only one spare.

Basic WDMs are passive units that require no electrical power connection. However, more elaborate types are available in configurations that allow the user to monitor individual channel power, alarm status, etc. These, of course, require a power connection.

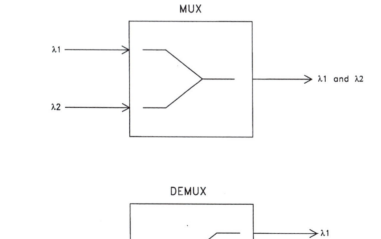

FIGURE 17.8   Unidirectional WDM.

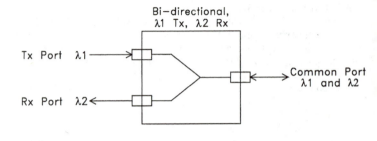

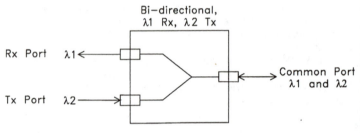

FIGURE 17.9   Bidirectional WDM.

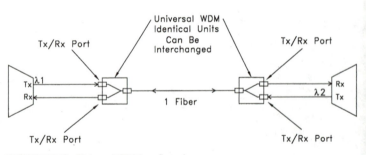

FIGURE 17.10   Universal WDM configuration.

WDMs are available in cassette-type enclosures or with fiber pigtails connecting each port (Fig. 17.11) Pigtail WDMs are smaller units but need to be spliced into the fiber span and are usually stored in a splice enclosure. The cassette-type WDMs are mounted into a shelf that holds a number of these units. Fiber patch cords are used to attach the cassettes to equipment. The cassette type of installation is preferred, since it allows for easier access and removal of a WDM in case it should fail.

WDMs add loss to a fiber link. This insertion loss is measured from the common port to any of the other transmit or receive ports. The loss must be included in all optical budget calculations. WDM insertion loss varies for dif-

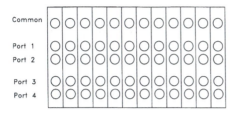

WDM as Cassettes in a Shelf

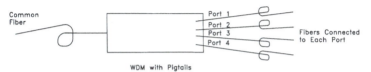

WDM with Pigtails

FIGURE 17.11   Pigtail four-channel WDM and a connectorized four-channel WDM.

ferent WDM types. The greater the number of channels for which the WDM is designed, the higher the WDM insertion loss.

17.1 WDM SPECIFICATIONS TO CONSIDER

The following WDM specifications should be considered when designing a fiber optic link for WDMs:

Specification	Unit	Description
Insertion loss	dB	Since two WDMs are used in a fiber link, this loss should be doubled to obtain the WDM loss for the fiber link
Number of channels	number	Number of wavelengths the WDM will couple into a fiber
Channel spectral bandwidth	nm	Ensure the laser sources will operate in the WDM channels
Isolation	dB	Minimum isolation that receiver equipment can tolerate between channels
PMD	ps/$\sqrt{\text{km}}$	Polarization mode dispersion added by the WDM
Return loss	dB	Reflectance caused by the unit

17.2 WDM APPLICATION'S

The following examples show a number of common WDM configurations:

17.2.1 Wide-Band 1310- and 1550-nm System

This type of configuration separates the fiber's 1310- and 1550-nm bands. It is often used to add capacity to an existing 1310-nm fiber system (see Fig. 7.12) as long as the existing system can tolerate an increased optical loss due to the insertion of these WDMs. The new communication system to be added must have transmitters using 1550-nm lasers (see Fig. 17.13).

A narrow-band WDM can be added to the wide band WDM system to further increase the fiber's capacity (see Fig. 17.14). Once again, the additional WDMs will increase the overall link optical loss.

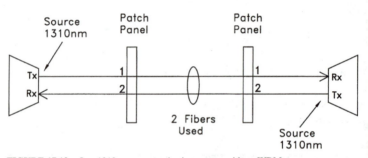

FIGURE 17.12   One 1310-nm communication system without WDMs.

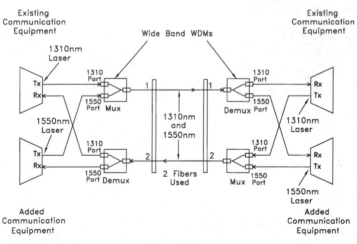

FIGURE 17.13   Two communication systems using wide-band WDMs.

17.2.2   Unidirectional WDM System

Unidirectional WDMs with 2 or more channels can be used in this configuration to achieve a high fiber capacity (see Fig. 17.15). However, at least two fibers are required—one to connect to the optical transmitters, and the other to connect to the receivers. For applications that do not require two-way communication, such as cable TV systems, only one fiber is necessary.

17.2.3   Bidirectional Narrow Band WDM System

The bidirectional WDM allows a communication system to use only one fiber for full duplex communication (see Fig. 17.16).

17.3   OPTICAL COUPLERS

Optical couplers connect three or more fibers. They physically split or combine optical light, and hence are sometimes referred to as splitters. A WDM is in fact a type of coupler that not only splits the light, but also is wavelength selective, directing certain wavelengths to certain fibers.

There are four types of couplers—T coupler, tree coupler (1 × N), star coupler, and wavelength-selective couplers (WDMs).

T Coupler.   T couplers, also referred to as taps, split one optical input into two outputs, or combine two inputs to one output. T couplers can split an optical signal equally or in a set ratio. An equal split would result in a drop of over 50% (3 dB) in optical power in each output. It is greater than 50%

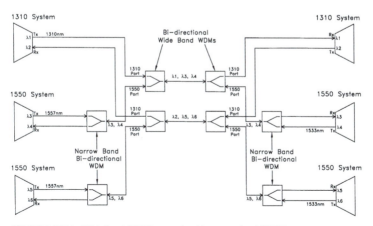

FIGURE 17.14   Wide-band WDMs cascade with narrow-band WDMs.

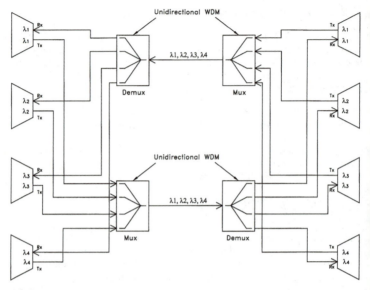

FIGURE 17.15 Unidirectional four-channel WDM system.

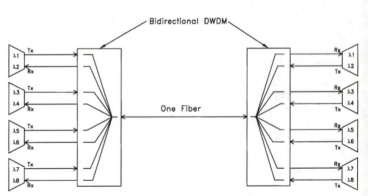

FIGURE 17.16 Bidirectional eight-channel DWDM.

because some light is inevitably lost due to the inefficiency of the coupler. If the light is split unequally, the coupler is referred to as a ratio splitter. For example, a 10/90 splitter would split 90% of optical power from the common port into the 90 port, and the other 10% of optical power into the 10 port (see Fig. 17.17).

typical dB conversion chart for ratio splitter follows:

Ratio	Approx. dB loss per port
Port A/Port B	Port A/Port B
50/50	3.5/3.5
45/55	4.0/3.0
40/60	4.5/2.6
35/65	5.2/2.3
33/67	5.4/2.1
30/70	5.8/1.9
25/75	6.7/1.7
20/80	7.6/1.3
15/85	9.0/1.0
10/90	11.0/0.8
5/95	14.2/0.5

A high ratio tap such as the 10/90 is often inserted into a fiber system so it will be able to continuously measure relative power levels or monitor optical signals with an optical spectrum analyzer.

Tree (1 × N) Coupler.   A tree coupler takes one optical signal from one fiber and splits it among multiple fiber outputs. Tree couplers are also available as multiple input combiners to one output fiber. Splitters are available as 1 × N units that split light equally from one fiber (common port) to N different fiber ports (see Fig. 17.18). The following chart shows the approximate loss for different 1 × N splitters that divide the light equally among all ports:

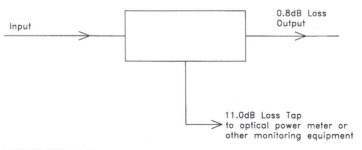

FIGURE 17.17   10/90 tap.

Number of splitter ports (N)	Approximate loss in dB per port	Number of splitter ports (N)	Approximate loss in dB per port
2	3.5	17	12.7
3	5.0	18	13.0
4	6.2	19	13.2
5	7.2	20	13.4
6	8.0	21	13.7
7	8.7	22	13.9
8	9.3	23	14.1
9	9.9	24	14.3
10	10.3	25	14.4
11	10.8	26	14.6
12	11.2	27	14.8
13	11.5	28	15.0
14	11.8	29	15.1
15	12.1	30	15.3
16	12.4	31	15.4

1 x N couplers are often used in the cable TV industry to distribute cable TV channel optical signals by fiber to multiple homes.

Star Coupler. Star couplers have multiple inputs and outputs, and frequently, an equal number of both. Star couplers take optical signals from all input fibers and distribute them amongst all output fibers (M × N) (see Fig 17.19)

17.4 OPTICAL SWITCH

An optical switch is a unit that can switch light from one fiber to another. It is available in an M × N configuration (M representing the number of input

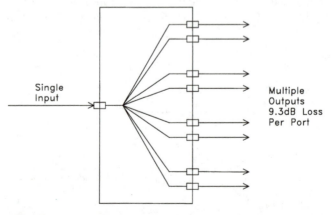

FIGURE 17.18   1 × N splitter.

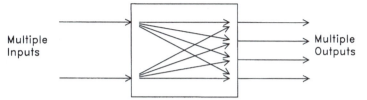

FIGURE 17.19 Star coupler.

fibers and N representing the number of output fibers) (see Fig. 17.20). This switch functions with either connectors or pigtails. It works by electrically activating an electromechanical device such as a stepper motor, which in turn moves a mirror, lens, or prism to redirect the light to another fiber.

An optical switch unit is fully bidirectional. (The input and output references are used just for convenience in order to identify the ports.) An important specification for the switch is its switch time, which is the time required for the unit to switch from one fiber to another. The switch time can, for some applications, be critical. A typical specification list for an optical switch follows:

Specification	Unit	Description
Switching time	ms	Time required to switch to a fiber
Insertion loss	dB	
Return loss	dB	Maximum level of reflections the unit caused by (reflectance)
Polarization-dependent loss	dB	
Repeatability	dB	Maximum change in loss over its lifetime
Cross-talk	dB	Optical interference caused by adjacent fibers
Fiber type		Single mode/multimode

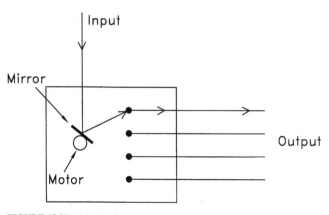

FIGURE 17.20 1 × 4 optical switch.

17.5   OPTICAL ATTENUATORS

Optical attenuators are passive units that decrease the optical power of an optical signal. They attenuate all wavelengths of optical light in the fiber.

Optical attenuators are available either as a fixed or variable configuration, both of which are common. The fixed attenuator (Pad) is purchased for a specified loss that cannot be altered. The variable optical attenuator (VOA) can have its loss adjusted over its specified range by turning a small set screw on the unit.

Pads are available with pigtails attached, to be used as fiber patch cords, or as a unit that mounts between the fiber jumper connector and bulkhead adapter (called a buildout or inline Pad) at the fiber distribution panel or lightwave equipment.

VOAs are usually larger than Pads and are therefore available with pigtails to be connected as a fiber jumper.

The most common use of an attenuator is to reduce optical power at the lightwave equipment receiver. Many receivers designed for long-haul applications are sensitive to optical overload. If the fiber optic link does not have sufficient loss, as specified by the manufacturer, then additional loss must be added to prevent

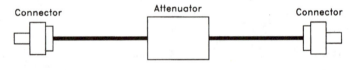

In Line Jumper Attenuator

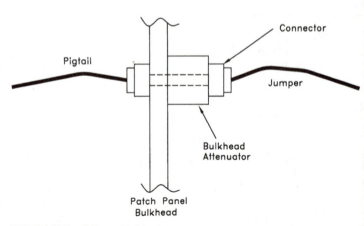

FIGURE 17.21   In-line and bulkhead attenuators.

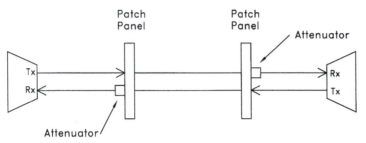

FIGURE 17.22 Proper and improper insertion of attenuators.

receiver overload. A Pad or VOA is installed at the receiver to reduce the received light level to within manufacturer's specifications. Attenuators should not be placed at the transmitter end of the fiber. This is because back reflections caused by the attenuator into the transmitter may affect transmitter operation. (see Fig. 17.22) Typical specifications for attenuators:

Specification	Unit	Description
Type		Variable or fixed
Mounting		Jumper style, in-line, bulkhead
Insertion loss	dB	This will be a range for a VOA
Return loss	dB	Maximum level of reflections (reflectance) caused by the unit
Polarization Dependent loss	dB	
Fiber type		Single mode/multimode

17.6 OPTICAL ISOLATORS

Optical isolators pass light in one direction with a low loss (less than 0.5 dB), but block light transmission in the opposite direction with high loss (40 to 70 dB). They are used to reduce back reflections from reaching sensitive laser transmitters. Excessive reflections can cause laser instability and transmission errors. Typical specifications for isolators:

Specification	unit
Mounting	Jumper or bulkhead
Insertion loss	dB
Isolation	dB
Polarization Dependent loss	dB
Wavelength range	nm
Fiber type	Single mode/multimode

CHAPTER 18
SONET/SDH

Synchronous Optical Network (SONET) is an American National Standards Institute (ANSI) standard for communications over optical fiber. Since its introduction in 1984, it has been in deployment by every major carrier in North America.

An equivalent transmission standard approved by International Telecommunication Union-Telecomm (ITU-T, formerly CCITT) is called Synchronous Digital Hierarchy (SDH). It has been accepted and used worldwide outside of North America.

A SONET terminal is a type of multiplexer that can multiplex many digital signals into a single optical channel. This is accomplished by a byte-interleaved multiplexing scheme. The base signal in SONET is called the synchronous transport signal level 1 (STS-1), which operates at 51.84 Mbps. Higher-level signals are formed by multiplexing together integer multiples of STS-1s. Then, once the highest-level aggregate is formed, the aggregate is converted to an optic signal (optical carrier—OC signal).

SONET and SDH levels are as follows:

SONET electrical	SONET optical	SDH	Line rate	Voice channels (64 kbps)	DS3s
STS-1	OC-1	STM-0	51.84 Mbps	672	1
STS-3	OC-3	STM-1	155.52 Mbps	2,016	3
STS-12	OC-12	STM-4	622.08 Mbps	8,064	12
STS-48	OC-48	STM-16	2488.32 Mbps	32,256	48
STS-192	OC-192	STM-64	9.95328 Gbps (10 Gbps)	129,024	192
STS-768	OC-768	STM-256	39.81312 Gbps (40 Gbps)	516,096	768

At the time of writing this book, the OC-768 (STM-256) was not commercially available.

SONET and SDH provide high-bandwidth and failure-resistant transmission technologies for networks. They also provide monitoring capabilities of all network elements and methods of restoring transmission loss due to fiber breaks quickly and effectively.

18.1   SONET ARCHITECTURE

One primary benefit in deploying SONET/SDH equipment is the ability of the system to recognize and reroute traffic in the event of a fiber cut or optical transmitter/receiver failure, or occurrence of significant degradation in a fiber. This survivability can be achieved in a SONET/SDH linear or ring architecture. Both architectures have their applications, advantages, and disadvantages.

18.1.1   Linear Architecture

A linear system is made up of two end terminals and a number of regenerators in between if required. Traffic flows from one terminal to another, without any means of adding or dropping traffic at intermediate sites (see Fig. 18.1).

This architecture requires four fibers to connect the two terminals. Two fibers are designated as transmit and receive Work fibers and carry the traffic under normal operating conditions.

The other two fibers are designated as Protect fibers. They do not carry traffic under normal conditions, but instead remain live. In the event of a fiber cut of any one of the work fibers, the SONET/SDH terminals recognize the event and switch all the traffic onto the Protect fibers. Switching occurs in 50 ms or less, so disruption to traffic is minimal.

Once the Work fibers are restored, the traffic is then switched back onto the Work fibers. This configuration also guards against laser transmitter or receiver card failure. If a transmitter card fails on one of the Work fibers, the system switches the traffic to the Protect fibers just like a fiber failure. Once the card is replaced, the system can restore the traffic onto the Work fibers.

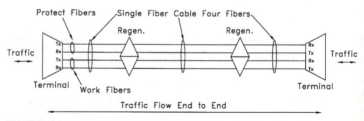

FIGURE 18.1   Linear system with traffic flowing from one terminal to another.

If a fiber or an optical card fails on the Protect fibers, then of course the system does not switch traffic but instead sends an alarm to the network manager identifying the problem so that it can be quickly repaired. This system always carries traffic on its Work fibers unless there is a degradation, fiber cut, or card failure of the Work fibers. The Protect fibers are live and continuously monitored by the system but do not carry traffic unless there is a failure in the Work fibers or cards.

In this system, four fibers can be used in one fiber optic cable. However, if that cable gets cut, then there would be a complete loss in traffic, since the system would not be able to switch the traffic to any good active fibers.

Alternatively, the system can be deployed in a physically diverse route (see Fig. 18.2). Two separate and physically diverse fiber cables can be used for the Work and Protect fibers. Since the two cables are physically diverse, a cut in one cable would not affect the second cable and would therefore not result in loss of traffic. Only a traffic switch or alarm would occur from the bad cable to the good cable.

This physically diverse fiber routing configuration for a linear SONET/SDH system provides the best protection of traffic between the two terminals. The main disadvantage of this architecture is that traffic cannot drop off at intermediate sites. All traffic must flow between the two end terminals.

18.1.2 Ring Architecture

A SONET ring system is made up of two or more SONET add/drop multiplexers (ADMs) that are connected in a ring by two or four fibers.

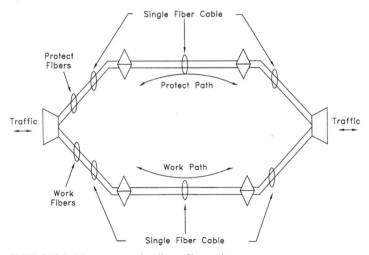

FIGURE 18.2   Linear system using diverse fiber routing.

Regenerators can be used at sites where traffic is to only pass through and not drop off. Unlike a linear system, the SONET ring system can have up to 16 ADMs in the ring with a capability of adding or dropping traffic at the ADM locations (see Fig. 18.3).

Two common types of ring configurations are the two-fiber bidirectional line switched ring (BLSR) and four-fiber BLSR ring.

Two-Fiber BLSR Ring.   The two-fiber BLSR ring configuration uses two fibers for communication around the ring. Each ADM has a total of four fiber connections, two fibers (transmit and receive) in one direction and two fibers in the other direction (see Fig. 18.4).

The capacity in any one direction out of an ADM is only half of its OC-n capacity. This is because half of the ring capacity is used for traffic, while the other half is reserved for protection in case of a fiber failure.

If traffic needs to flow between NE1 and NE2, a connection is established using span 1 fibers. If other traffic needs to be routed between NE1 and NE3, a connection can be established using spans 3 and 4, and NE4 (see Fig. 18.4).

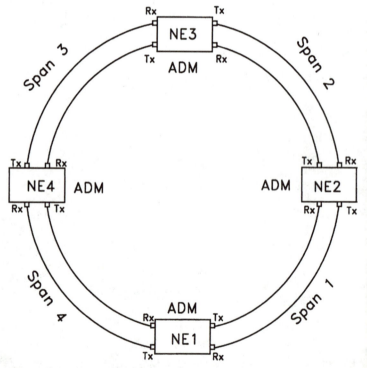

FIGURE 18.3   Two-fiber BLSR ring.

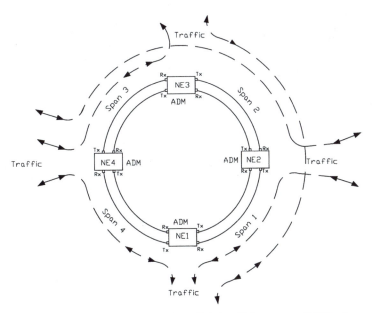

FIGURE 18.4 A two-fiber SONET ring with traffic being added at numerous ADM locations.

If there is a failure in a span, then all traffic through that span is switched and rerouted around the ring using the protection half of the OC-n capacity. For example, in an OC-48 two-fiber BLSR ring, a maximum of 24 DS3s carrying traffic can be routed in each direction out of an ADM, around the ring (see Fig. 18.5). (Note that an OC-1 can carry one DS3.)

However, the OC-48 total capacity is 48 DS3s around the ring. The other 24 DS3 channels around the ring are reserved for protection traffic only, to be used if there is a failure in one of the spans.

In the event of a cable cut, fiber problem, or optics failure in one of the spans, such as span 1, for example, then NEs 1 and 2 will switch the traffic that was on span 1 onto spans 2, 3 and 4. By using the reserved protection DS3 channels, the traffic will be rerouted around the ring, and, as a result, bypass the failed span (see Fig. 18.6).

When the failed span is restored, then the traffic will be switched back onto span 1 fibers. During this failure, traffic flowing between NEs 1 and 3 using spans 3 and 4 is unaffected.

Four-Fiber BLSR Ring. In a four-fiber BLSR ring configuration, each ADM communicates through four fibers, in each direction, around the ring. This comes to a total of eight fiber connections for each ADM (see Fig. 18.7). Two fibers carry all traffic, while the other two fibers act as Protect fibers and remain ready in case of a failure.

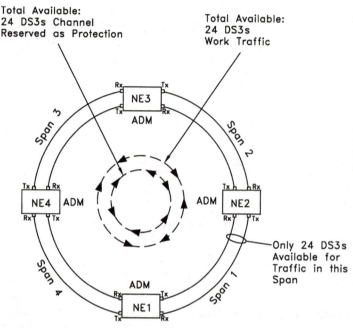

FIGURE 18.5 OC-48 two-fiber BLSR ring.

For example, an OC-48 four-fiber BLSR ring can carry 48 active DS3s around the ring. At an ADM, 48 DS3s can enter from both fiber directions for a possible maximum of 96 DS3s at the one ADM (dependent on equipment). See Fig. 18.8.

The four-fiber BLSR ring protects against failure in two ways—span switching and line switching.

A span switch would occur in a span if there were a failure with the Work fibers or Work transmit/receive cards. The traffic in the span Work fibers would be switched to the same span Protect fibers. Traffic would remain in the span, but just on different fibers. Other parts of the ring would not be affected (see Fig. 18.9). The ring can even survive multiple span failures without losing traffic (see Fig. 18.10).

Often, the four fibers in a span are in the same cable. Therefore, if the cable gets cut, the ring will perform a line switch. All traffic through the failed span is rerouted around the ring, onto the Protect fibers (see Fig. 18.11).

Both two- and four-fiber BLSR rings are commonly deployed. Their advantages and disadvantages for the application should be carefully considered.

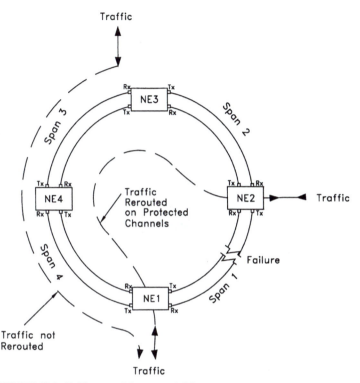

FIGURE 18.6 Traffic rerouted due to span 1 failure.

Two-Fiber BLSR

Advantages:

- Less expensive due to fewer required optical cards
- Less fibers required

Disadvantages:

- Can only carry half the OC-n capacity for traffic in either direction out of an ADM because the other half is reserved for protection

Common applications:

- Metropolitan or local areas, deployed by local exchange carriers (LECs)

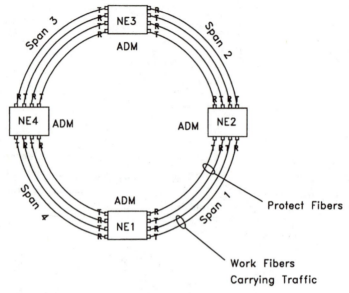

FIGURE 18.7   Four-fiber BLSR ring.

Four-Fiber BLSR
Advantages:

- Can carry the full OC-n traffic around the ring
- Provides line and span switching
- Can handle multiple span failures

Disadvantages:

- More expensive
- Twice as many fibers required
- More complex

Common applications:

- Long-haul traffic, regional or national areas, deployed for interexchange carriers (IXCs)

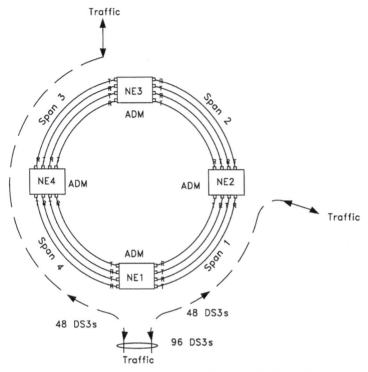

FIGURE 18.8 Maximum DS3 traffic for an OC-48 four-fiber BLSR ring ADM.

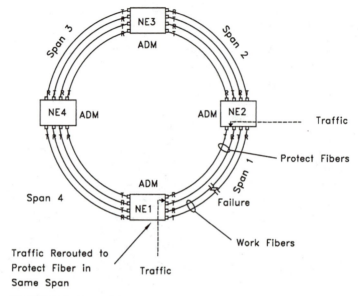

FIGURE 18.9 Span switch.

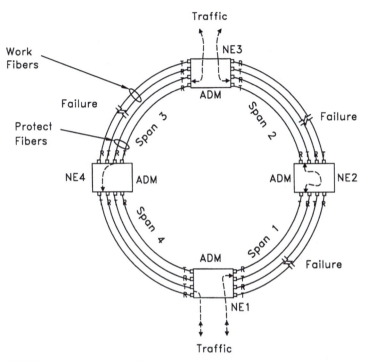

FIGURE 18.10 Multiple span failures.

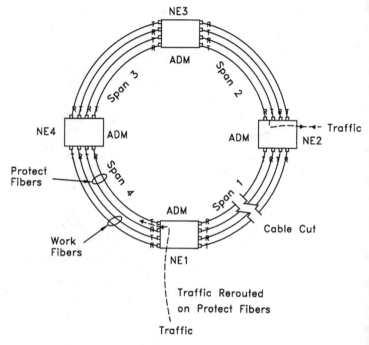

FIGURE 18.11 Line switch.

CHAPTER 19
LAN

19.1 ETHERNET/IP OVER FIBER

Fiber deployment in Ethernet/IP networks is becoming very popular. There is some consensus in the industry that fiber will eventually displace copper as the choice medium for deploying LAN systems. However, the time frame for this copper-to-fiber transition is not known. Today, the amount of fiber deployed in LAN systems is only a small fraction of the copper installations.

Even though fiber has some advantages over copper (see Chap. 1), the higher fiber cable cost, relative to copper, and the additional cost of the fiber-to-copper conversion makes it impractical for many installations. However, as data rates increase from 10 to 100 Mbps, and 100 Mbps to 1 Gbps or even 10 Gbps, copper may not be able to support these higher rates, making the conversion to fiber inevitable.

Standards for the deployment of optical fiber for Ethernet communications have either been developed or are now under development by IEEE, TIA/EIA, and other organizations.

The existing Ethernet LAN standards are summarized in the table below:

Data rate (Mbps)	Copper standard	Fiber standard	Wavelength band (nm)
10	10Base-T	10Base-FL	850
100	100Base-TX	100Base-FX	1310
100	100Base-TX	100Base-SX	850
1000	1000Base-TX	1000Base-LX	1310
1000	1000Base-TX	1000Base-SX	850

The 1550-nm band is not presently standardized for Ethernet traffic. The table below shows the relationship between bandwidths and available optical technology:

Wavelength band	Lower limit	Center	Upper limit	Source	Detector	Fiber type
850 nm	820 nm	850 nm	920 nm	LED/VCSEL	PIN	Multimode
1310 nm	1270 nm	1300 nm	1380 nm	LED/VCSEL/ laser	PIN/APD	Multimode/ single mode

The optical fiber long wavelength upper limit at 1380 nm is primarily due to water ion (OH) absorption. The utilization of new, specialized fibers has significantly reduced this barrier and allowed for continuous transmission from 1270 to 1600 nm. The spectral band limits in fiber are expanding due to improvements in the fiber and availability of lasers that can utilize these new bands.

Generally, the short-wavelength (850 nm) optical electronics are much less expensive than similar longer-wave (1310 nm) electronics. If the optical budgets are sufficient, short-wavelength electronics are preferable. Alternately, long-wavelength communication is deployed at usually a higher cost but will allow for longer transmission distance.

Transmission distance is dependent on the transmission rate (10/100/1000 Mbps) and the type of fiber. The IEEE 802.3 standard specifies maximum distances for different data rates. The IEEE 802.3u standard defines the use for fiber for 100-Mbps application, and IEEE 802.3z standard for 1000-Mbps.

19.2   LAN CABLE INSTALLATION

Installation of fiber cable in commercial buildings can be described in six subsections:

1. Horizontal Cabling.   Horizontal cabling is the cabling section that extends from the user's work area (WA) wall outlet to the telecommunication closet horizontal cross-connect (HC) (see Fig. 19.1). It also includes any fiber jumpers used in the telecommunication closet. The cabling is configured as a star topology with the telecommunication closet as the hub. Horizontal cabling is often placed above suspended ceilings or under raised floors. A tight, buffered cable is a good choice for this type of cable for two reasons. First, it can be terminated with connectors without the need to use splice trays. Second, it can be purchased preterminated, with connectors at both ends. The cable is a small size, at usually two or four fibers dedicated for each user.

2. Backbone Cable.   Backbone cabling (see Fig. 19.2) provides the fiber connection between the telecommunication closet horizontal cross-connect

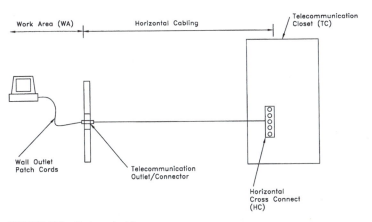

FIGURE 19.1 Horizontal cabling.

and intermediate cross-connect (IC) or main cross connect (MC) equipment rooms or cable entrance facilities. This cabling, as with the horizontal cabling, is also configured as a star topology. Building backbone cabling is installed into building risers, metallic conduit, or cable tray systems. This cable can be tight buffer or loose tube. Conduit can be used to provide the cable with additional protection and isolation from the local environment. Campus backbone cabling runs in between buildings and can be directly buried or placed into ducts. (see Fig. 19.2)

Loose-tube type cable is used for its ruggedness and high fiber counts. The backbone cable fiber count is sized larger in order to accommodate present individual fibers and future expansion. Sufficient fibers should be available to interconnect with all horizontal cable fibers in the telecommunications closet and extras for future growth. Backbone cable is much more difficult to install than horizontal cable; therefore, careful planning of this cable section is recommended. Large cable fiber counts such as 72, 144, or more fibers can be considered.

3. Work Area Cable.   The work area (wa) cable interconnects the horizontal cabling at the wall outlet to the desktop computer. This cable is usually a short simplex or duplex fiber jumper.

4. Telecommunication Closet.   The telecommunication closet (TC) is a room in the building used for horizontal cross-connects or intermediate cross-connects. It provides interconnection between backbone cabling and horizontal cabling, two or more backbone cables, or entrance cable to backbone cable.

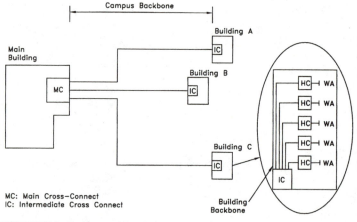

FIGURE 19.2 Backbone cabling.

5. Equipment Room. The equipment room is the room in which the telecommunications equipment is located. All the backbone cables terminate here, and this is the common point of the star topology. This is normally the location of the main cross-connect fiber distribution panel.

6. Entrance Facility. The entrance facility is where outside fiber cables enter the building and connect with the building's backbone cables.

19.3 10/100-MBPS LAN

For LAN systems operating at 10 or 100 Mbps, fiber cable lengths for installation should be obtained from the most recent TIA/EIA 568A and IEEE 802.3u standards documents. The following chart suggests maximum lengths for the given LAN type:

Bandwidth (MHz*km)   Max. length (meters)

LAN Type	Wavelength (nm)	50 μm	62.5 μm	50 μm	62.5 μm
10Base-F	850	500	160	1000	2000
100Base-FX	1300	500	500	2000	2000
100Base-SX	850		500		300

Note: Maximum fiber loss for fiber at 850 nm is 3.75 dB/km and for 1310 nm is 1.5 dB/km.

The above maximum fiber lengths are total lengths, including backbone cable length and horizontal cable length (see Fig. 19.3).

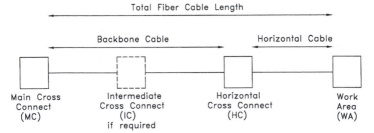

FIGURE 19.3   10-Mbps LAN cabling lengths.

19.4   1000-MBPS LAN

The IEEE 802.3z standard, also called Gigabit Ethernet, operates at 1.25 Gbps (see Fig.19.4). Gigabit Ethernet transmission equipment uses VCSEL or laser sources because LED sources are not able to operate at these high transmission rates. Lasers used with multimode fiber require a special offset launch patch cord (see Chap. 3) to counter the differential mode delay (DMD) affect.

The chart below shows typical maximum distances between LAN equipment, including backbone cabling and horizontal cabling. Exact distances should be obtained from the latest-mentioned specifications.

1000Base-SX for 850-nm Light Source Maximum Transmission Distance

Fiber core size (μm)	Fiber modal bandwidth (MHz*km)	Maximum cable length (m)	Max loss (dB)
62.5	160	220	2.3
62.5	200	275	2.6
50	400	500	3.3
50	500	550	3.5

1000Base-LX for 1310-nm Light Source Maximum Transmission Distance

Fiber core size (μm)	Fiber modal bandwidth (MHz*km)	Maximum cable length (m)	Max loss (dB)
62.5	500	550	2.3
50	500	550	2.3
50	400	550	2.3
9 SM	Not applicable	5000	4.5

Notes:

1. The maximum loss in the above two charts assumes the measurement is done using a laser source.

2. The maximum cable lengths for multimode fiber assume that an offset launch patch cord is used (see Chap. 3) to counter DMD delay.

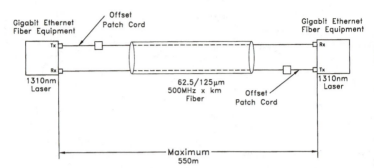

FIGURE 19.4   Gigabit Ethernet installation using 62.5/125-μm fiber.

These maximum distances may be reduced if high losses are encountered in the fiber link due to factors such as dirty connectors, too many connectors, high splice loss, improperly installed fiber cable, or poor equipment transmitter/receiver optics. An optical power meter reading should always be taken to confirm total link loss and ensure it is within equipment specifications.

19.5   OTHER LAN STANDARD CABLE LENGTHS

The following chart suggests maximum fiber cable lengths for other LAN types:

LAN	Data rate	Wavelength	Bandwidth (MHz*km)		Max. length (ms)	
type	(Mbps)	(nm)	50 μm	62.5 μm	50 μm	62.5 μm
Token Ring	16	850	500	160	1000	2000
FDDI PMD	100	1300	500	500	2000	2000
FDDI LCF	100	1300	500	500	500	500
Fibre Channel	1063	850	500	160	500	175
Fibre Channel	531	850	500	160	1500	350
ATM	155	850	500	160	2000	2000
ATM	155	1300	500	500	1000	1000
ATM	622	850	500	160	300	300
ATM	622	1300	500	500	500	500
ATM	2500	1300	500	500	300	100

CHAPTER 20
FIBER SYSTEM DEPLOYMENT

20.1 OFFICE ENVIRONMENT

Figure 20.1 shows a possible office building installation. Lightwave equipment in buildings A and B is located in a secure telecommunication closet. The equipment is mounted in enclosed cabinets or in simple 19-in. mounting open equipment racks. Rack sizes of 23 and 24 in. are also common for larger equipment. Cabinets provide for a totally enclosed, neater-looking solution and are often used in showroom computer room installations. They should have at least a front and back door to provide adequate access to equipment and cabling. A smoked glass front door will allow the equipment's light status to be monitored without opening the door. Cabinet top and bottom can be open to allow for cable entrance and heat ventilation. In a busy room, these doors can be equipped with key locks to provide additional security.

Open-style equipment racks allow for better access to mounted equipment and cables. These racks are used in locked telecommunication closets that are dedicated for this purpose. Access is permitted to only authorized employees.

An example of an office installation is shown in Fig. 20.1 (building A). The center rack is assigned to lightwave equipment installation and the right rack is dedicated to fiber optic cable patch panels. Lightwave equipment and patch panels are assigned to separate racks, which is common for a large installation that requires plenty of room for expansion.

For smaller installations, both lightwave equipment and patch panels can be mounted in the same rack. Figure 20.2 shows a typical rack/cabinet layout with lightwave equipment and the patch panel mounted together.

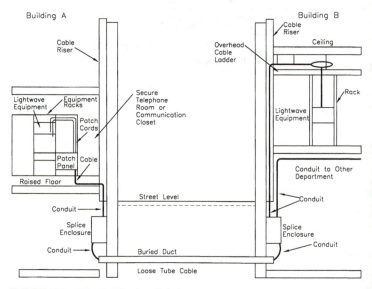

FIGURE 20.1   Office building installation.

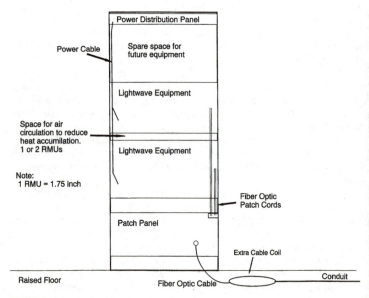

FIGURE 20.2   Rack/cabinet layout.

In building A in Fig. 20.1, two fiber optic patch cords connect the lightwave equipment to the patch panel mounted in the adjacent rack. Most lightwave equipment uses two optical fibers for full duplex communication— one fiber for the transmit signal and the other fiber for the receive signal. Lightwave equipment that is capable of full duplex transmission on one fiber is available by using WDMs. Computer equipment located in work areas connects to the telecommunication closet and lightwave equipment by horizontal cabling. Horizontal cabling can be fiber optic or copper cable.

A fire-rated, tight-buffered indoor fiber optic cable runs in a conduit between the patch panel and the basement splice enclosure. An oversized conduit can be installed to accommodate future cables. All conduit bends must be smooth, with a radius larger than the fiber optic cable's minimum bending radius for the loaded condition.

The tight-buffered cable is spliced to the outdoor loose-tube cable in the basement splice enclosure. The splice enclosure has the capacity to accommodate a number of cables to allow the fiber to be branched to different locations in the building. A patch panel can also be used in this location instead of the splice enclosure. The patch panel will provide a fiber intermediate cross-connect facility and allow for easy future changes.

The outdoor loose-tube fiber optic cable is routed through buried ducts into building B. The loose-tube cable is spliced to two tight-buffered cables servicing two different departments. A third outdoor loose-tube cable could also have easily been spliced here to continue the cable link to a third building. Again, a patch panel could have been used instead to provide an intermediate cross-connect facility.

The tight-buffered cable continues to an upper floor through conduits in the building's cable riser. It is then routed along a suspended ceiling to a temporary equipment installation. An additional length of coiled cable is placed in the ceiling to allow the equipment to be moved to a permanent telecommunication closet sometime in the future. The amount of cable in the coil depends on the distance of the route to the permanent installation as well as the amount of cable needed for splicing in a patch panel. The entire cable length to the cable coil is protected by conduits.

The cable drops from the ceiling to the lightwave equipment along a supporting pole. Because of the temporary nature of the installation, a patch panel is not used. Connectors are installed directly onto the cable's buffered optical fiber.

20.2 *INDUSTRIAL PLANT INSTALLATION*

Heavy industrial environments require special consideration, and all components should be designed to meet the harsher conditions. Fiber optic cable routing throughout a plant should always be protected by conduits or heavy armored jackets. Cable temperature specifications should exceed all anticipated environmental variations. Outdoor cable installations may require additional protection from the environment or from the movement of heavy equipment. Single-armor or double-armor cables with heavy polyethylene jackets should be considered.

As Fig. 20.3 shows, fiber optic cable can be used to provide communications to a remote plant. Aerial cable installations are common because of the available electrical pole line route. The fiber optic cable is placed as high as possible (in accordance with all electrical codes) on the pole to provide maximum ground clearance. At locations where high-load road traffic is anticipated, the cable can be buried in a duct. A 10-ft heavy steel guard or conduit can be placed around the cable at the pole riser to protect the cable from accidental damage by vehicles. The cable is routed in conduits and cable trays to the electrical rooms at both locations. Here the cable is terminated in a dust-free enclosure or in other enclosures properly rated for such installations.

Figure 20.4 shows an example of an industrial fiber optic cabinet layout. The cabinet is used to house fiber cable terminations and lightwave equipment and store excess fiber optic cable. Equipment is wall-mounted in the cabinet. A typical size is 1.8 m tall by 0.6 or 0.9 m wide and 0.45 m deep (6 ft tall by 2 or 3 ft wide and 18 in. deep). Construction is to industry code, usually consisting of heavy galvanized steel with a solid-hinged sealing door. A door lock is often used to discourage unauthorized entry.

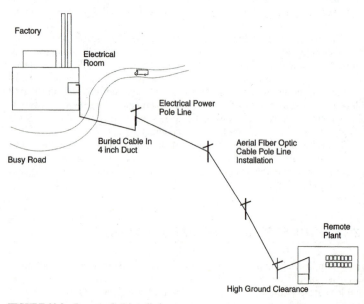

FIGURE 20.3   Remote plant installation.

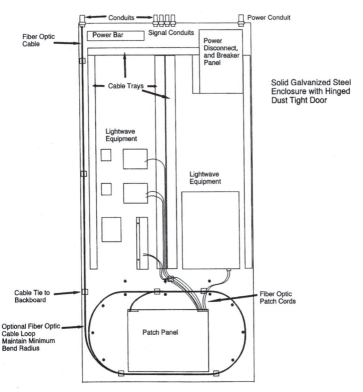

FIGURE 20.4 Industrial fiber optic cabinet layout.

The cabinet's equipment layout allows room for patch panels, terminating equipment, power supplies, electrical power disconnects and breaker panels, the test equipment's electrical power bar, and various wiring troughs. Storage of extra fiber optic cable can be accommodated in a cable loop at the bottom of the cabinet. Cable loop minimum bending radius (for static installation) should always be observed. Cable conduits enter the cabinet from the top or bottom and are properly sealed.

Fiber optic patch cords can be routed in dedicated cable trays to lightwave equipment. Lightwave equipment is the wall-mounted type and is secured to the cabinet's back panel. Additional space for associated lightwave equipment and power supplies may be required. Power wiring and signal wiring are placed in separate cable trays and routed to appropriate equipment in other cabinets. Center cable trays can be eliminated in order to mount large equipment.

20.3   ETHERNET

Fiber optics can be used to extend Ethernet LAN distances further than with conventional cabling such as 10Base5, 10Base2, or 10BaseT. Coaxial, 10Base5 cabling is limited to 500 m (1640 ft). Fiber optic modems (media converters) need to be used to convert the electrical LAN signal and back. This type of deployment has distance and fiber-type restrictions. Fiber optic modem manual should be consulted. Fiber-type 62.5/125 μm and ST connectors are common for these installations. The use of fiber optic bridges can also extend the distance to over 30 km on single mode fiber (see Fig. 20.5). See Chap. 19 for more details on LAN networks.

20.4   FDDI

Fiber distributed data interface (FDDI) is a standard for a high-data-rate (100-Mbps) area networking technology that uses fiber optics as the transmission medium. The system is configured in a dual ring topology (similar to a token ring), using a pair of optical fibers to connect each device onto the ring. It is often used as a high-speed campus backbone, a gateway to lower-data-rate LANs, or a connector between mainframe computers. As a data highway, FDDI has plenty of bandwidth to carry large amounts of data and is robust enough to provide good fault tolerance and redundancy (see Fig. 20.6).

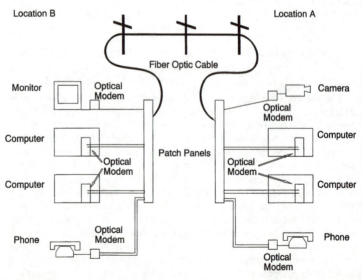

FIGURE 20.5   Ethernet system.

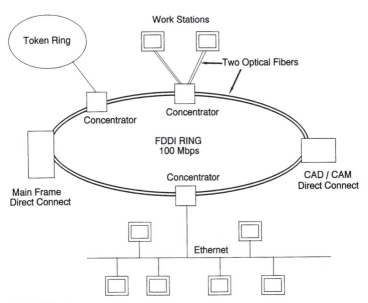

FIGURE 20.6 FDDI backbone network.

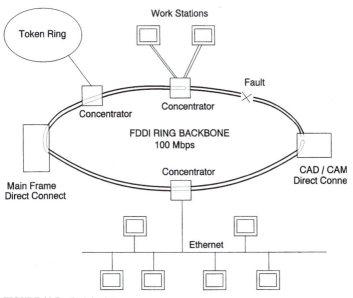

FIGURE 20.7 Fault in ring.

This dual ring architecture provides a high degree of reliability and security. Under normal operating conditions, the primary ring is used to carry data, while the secondary ring remains idle. In the event of a fault, a break in the fiber optic cable, or a device failure, stations on both sides of the fault detect and automatically loop back the primary fiber onto the secondary optical fiber. This bypasses the fault condition and recovers the ring operation (see Fig. 20.7).

Each station on the ring can be up to 2 km apart when using FDDI-grade optical fiber. The system can support up to 500 stations on the ring. Maximum ring length is 100 km (according to FDDI standards). New proprietary interfaces using single-mode fiber can increase these lengths, but individual product manufacturers should be consulted. FDDI standard fiber used for installation is multimode 62.5/125-μm FDDI grade operating at 1310 nm. Standard optical fiber connectors are the dual-fiber, keyed FDDI type. Some equipment can also accept other connectors such as the ST, FC, or SC type.

Communication occurs by timed token-passing protocol. This eliminates any delays that could be caused by packet collisions, as occur in Ethernet.

CHAPTER 21
MAINTENANCE

Fiber optic cable systems are generally maintenance free. Once installed properly, they will operate properly for many years, if not disturbed. The key is to ensure that the cable system is not disturbed. This is much easier said than done. Construction is ongoing, year round, and there is always a possibility that a cable may be hit accidentally. In addition to the human factor, animals such as rodents and birds often take a liking to cable and peck or chew at it relentlessly. Other environmental damage can occur from lightning strikes or heavy wind and ice storms. Therefore, even though a fiber manufacturer may claim that their cabling system is maintenance free, it may be best to schedule periodic inspection routines.

21.1 NON-SERVICE-AFFECTING MAINTENANCE

The fiber optic cable system should be visually inspected, at least annually. Patch cord and cable bends should be checked to ensure that the minimum bending radius is not compromised.

Patch cords should be neatly stored or secured in cable trays. Do not tie wrap or bend too tightly. They should never be left dangling.

Fiber optic cable bends should be checked to ensure that they are not too tight. Cable jackets can be inspected for damage, such as cuts, tears, or deformations.

Lightwave equipment and patch panel locations should be checked for cleanliness. Cabinet doors should remain closed at all times. Dust should not be permitted to enter or accumulate on equipment. In dusty environments, special sealed cabinets should be used.

Optical attenuation in nonoperational optical fibers (spare fibers) can be measured using a light source and power meter and can be compared to installation records to determine optical fiber deterioration. Under normal operating conditions, fiber attenuation should remain constant over many years.

Lightwave equipment lasers age over time, which results in the decrease of their optical output level. This level can be checked annually. Some types of lightwave equipment provide the means for an operator to check this optical level without disconnecting fibers and affecting equipment operation (consult the manufacturer's operations manual). Manufacturer's specifications should be consulted to determine proper laser output levels.

Pole line installations should be visually inspected for any damage to the cable, messenger, or supporting structure. Messenger and fiber optic cable sag should also be checked. Overhead cables are susceptible to environmental damage, such as wind or ice, as well as damage from birds, rodents, humans, or gunfire.

Underground cable vaults can be checked for cable and support integrity and for rusting. Cable damage can occur from many sources: installation of other parties' cables; damage by rodents, water, ice, and fire; and cable support corrosion.

All duct, innerduct, and conduit seals can be checked for integrity. Water should not be allowed to enter any fiber optic duct, innerduct, or conduit assembly.

The entire buried cable plant route should be inspected. Be on the lookout for any activity that may disturb or damage the cable, such as new road construction.

21.2 SERVICE-AFFECTING MAINTENANCE

Service-affecting maintenance can be performed as frequently as recommended by the manufacturer. Normally, equipment is stable for long periods.

Operational fibers can be disconnected and fiber attenuation measured using OTDR and power source/meter. Results are then compared to records to determine any increase in attenuation. Lightwave equipment's optical output power can be measured and compared to records to determine laser or LED aging.

Optical fiber's reflected power can be measured to ensure stable laser operation (single-mode fiber only). Receiver optical threshold and BER can also be checked and compared to records.

CHAPTER 22
EMERGENCY CABLE REPAIR

Emergency cable repair should be carefully planned to prepare for a quick and efficient cable restoration. Because of the very high information capacity of a fiber optic cable, a broken cable can affect thousands of telephone calls and data circuits. Restoration time is usually of great importance. The plan should include sufficient emergency repair supplies and test and repair equipment, as well as trained repair personnel who can be available at a moment's notice.

Equipment and Supplies:

1. Sufficient lengths of repair cable or cables (500 m usually will do for most repairs). Also, when numerous cable reels are located strategically throughout the fiber network, access to a repair cable can be made quickly.

2. Proper fusion splicer and associated supplies. Cable stripping and cleaning supplies.

3. OTDR, power meter, and light source. Having two OTDRs ready will allow for quick splice-loss testing from both cable ends.

4. Gas-powered generator, field lights (for night repairs), and water pump (in case trench is flooded).

5. Sturdy splicing table, chair, and splicing van or tent.

6. Existing network cable and splicing diagrams and drawings.

7. Two outdoor splice enclosures (and hand holes for buried cables).

8. Mechanical splices for possible temporary repair.

9. Proper equipment is required when installing the new section of cable and removing the old section. For buried cable, proper trenching equipment such as a backhoe and operator is required.

10. Cable locating equipment, for buried cables.

Usually, sufficient slack is not available in an installed cable to perform a splice without the addition of a patch cable length. A patch cable will introduce two splice losses, as well as patch cable attenuation, into the link budget. The system link budget should be reviewed to ensure that splicing with a patch cable is possible. If it is not, replacement of the entire cable may be required (see Fig. 22.1).

Once the damaged section of cable is identified, it should be cut out. If immediate fiber optic cable restoration is required, a short length of a rugged cable can be laid on the ground and temporarily spliced (mechanical splices can be considered if the link budget is adequate and temperatures do not fall below freezing). A proper rugged-style cable can withstand such abuse as vehicle crushing, tangling, freezing in ice, and so on.

Method

1. Determine the exact location of the fiber break and the proper length of patch cable required for splicing at ground level. For outdoor repair, splicing is done preferably in a covered enclosure, such as a van or tent.

2. Determine the locations of splice enclosures. For outdoor repairs, splice enclosures should be outdoor and weather resistant, with a watertight seal. For pole line repairs, splice enclosures designed for mounting on a messenger or on the pole should be used.

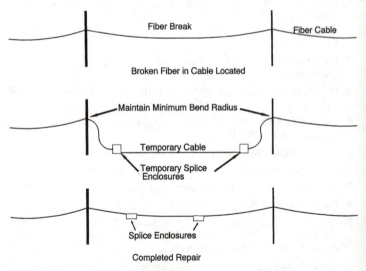

FIGURE 22.1   Aerial fiber optic cable repair.

3. Ensure that all repair hardware—including OTDR, power meter and source, fiber cleaning supplies, cable stripping supplies, and mechanical or fusion splicers—is available on site.
4. Ensure that all communication in the cable is terminated. In the event that only a few fibers in the cable are broken, the repair of the cable should be attempted only when traffic on all good fibers has been terminated.
5. Disconnect all fiber optic light sources that are connected to the cable needing repair.
6. Prepare the area around the damaged cable. For buried cable, the damaged cable should be carefully excavated. The replacement patch cable is placed alongside the damaged cable.
7. Once the splice enclosures and splicing location have been prepared, cut out the damaged section of cable. Using standard fusion splicing procedures, splice in the new patch cable. In the event that a fusion splicer is not available at the time of the repair, a temporary mechanical splice can be made to restore service quickly. Mechanical splices can then be replaced by fusion splices later when a fusion splicer and crew become available.
8. Test all splices with OTDR and ensure the splices are within acceptable loss (less than 0.1 dB). Test the complete fiber length with OTDR to also ensure that all damaged cable has been repaired. Use a power meter and source to measure total fiber loss and confirm that it is within the lightwave equipment optical loss budget.
9. After cable splicing and testing are complete, mount the splice enclosures to their permanent locations. For a pole line repair, the splice enclosures can be mounted on the messenger or on the pole (see Figs. 22.2 and 22.3). Store cable slack in a folded-over manner on the messenger, or in a coil on the pole, making sure to observe the cable's minimum bending radius. Also make sure that cable twisting does not occur. For buried cable repair, place hand holes and install splice enclosures in hand holes. Carefully coil excess cable in hand holes for possible future re-entry. Bury cable and handholes, and restore the surface to original condition.

Figure 22.2 shows a typical pole line splice enclosure installation on a messenger cable. Excess fiber optic cable length is stored folded over on the messenger. Minimum fiber optic cable bending radius should not be compromised. Splice enclosures are weatherproof and watertight and have special mounting brackets for mounting the messenger.

If mounting on the messenger is not desired, splice enclosures can also be pole mounted. Figure 22.3 shows a typical pole mount configuration. The fiber optic cable coil is placed directly above the splice enclosure. Make sure the cable's minimum bending radius for unloaded conditions is not compromised. Pole-mounting containers that will house both the cable coil and the splice enclosure are available. Enclosures can be mounted high to discourage unauthorized entry.

Figure 22.4 shows a typical handhole and splice enclosure installation.

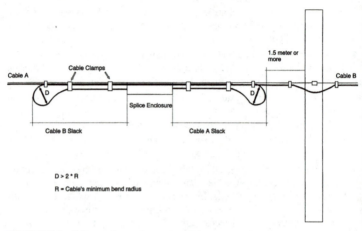

FIGURE 22.2 Splice enclosure mounted on messenger.

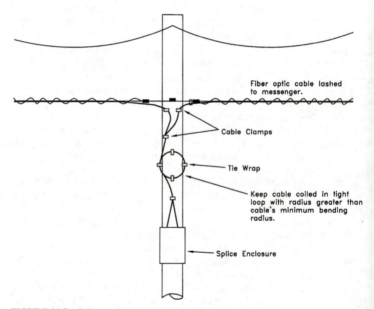

FIGURE 22.3 Splice enclosure mounted on pole.

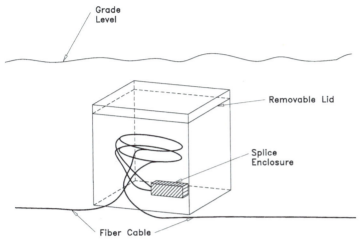

FIGURE 22.4 Handhole and splice enclosure installation.

CHAPTER 23
RECORDS

Accurate installation and maintenance records should be kept for future repair or system modifications.

Installation Records

1. Installed optical fiber OTDR test documentation should include:

- Fiber attenuation
- Individual fiber traces for complete fiber length
- Paper or computer disk records of all traces
- Losses of individual splices and connectors
- Losses of other anomalies
- Wavelengths tested and measurement directions
- Manufacturer, model, and serial number of the test equipment

2. Installed power meter test documentation should include:

- Total link loss for each fiber
- Source optical output power using standard transmission pattern, such as all ones
- Receiver optical threshold level
- Optical margin level
- Return loss measurement (laser sources only)
- Wavelengths tested and measurement directions
- Manufacturer, model, and serial number of test equipment

3. As-installed drawings should include:

- Fiber optic cable route drawings and details
- Splicing locations
- Optical fiber connection diagram
- Optical fiber assignments at patch panels and terminations
- Connector type
- OPptical fiber assignments at splice locations

- Installed fiber optic cable length
- Date of installation
- Installation crew members

4. Fiber optic cable details should include:

- Manufacturer of fiber optic cable
- Cable type, diameter, and weight
- Jacket type, fire rating, indoor or outdoor
- Manufacturer of optical fiber
- Optical fiber type
- Optical fiber core and cladding diameter
- Optical fiber attenuation per kilometer
- Optical fiber bandwidth and dispersion details

5. Lightwave equipment specifications should include:

- Manufacturer, model, and serial number
- Detailed operation, maintenance, and repair instructions
- Lightwave specifications, output power, receiver sensitivity, source type, operating wavelength, and so on
- Other equipment details

Maintenance Records

Maintenance records should include:

1. Date, crew members, and test equipment used to test and inspect fiber optic system.

2. Power meter and OTDR testing completed as shown in installation records and readings compared to installation records. These tests will require fibers to be taken out of service.

3. Cable route inspected and notes on condition of cable and support structure.

4. Patch panel and splicing assignments checked.

5. Detailed notes on any problems encountered or changes made and manufacturer model and serial number of all equipment tested or changed.

CHAPTER 24
TROUBLESHOOTING

A methodical, step-by-step approach should be used to isolate a problem. If the system is operational, all users should be informed of possible disturbances or outages. A log should be kept noting symptoms of the problem and the steps used to identify and rectify it.

Non-Service-Interrupting

1. Visually inspect all patch cords and patch panels to ensure that the cabling is correct and connected as required.
2. Check to ensure that all equipment has power and is in full, proper operational mode. Check all visual hardware and software indicators for proper configuration.
3. Ensure that patch cords and cables have not been bent tighter than their minimum bending radius.
4. Visually inspect the entire cable route for damage, tight bends, or abnormal conditions.
5. Ensure that all connections are secure. If the problem is intermittent, gently wiggle the cables and connectors to see if the problem occurs (this is possibly service interrupting). Some connectors such as the biconic type are susceptible to vibrations.

Service-Interrupting

1. Disconnect all patch cords and properly clean all connectors.
2. Test all patch cords. Replace if any are doubtful.
3. Measure attenuation of all fibers with a light source and power meter and compare to records.
4. Test all fibers with OTDR and compare results to records.

5. Connect power meter directly to the lightwave equipment source and measure optical output power. Compare to records.

6. Connect power meter at the receiver location, facing the optical fiber, and measure the optical power of the distant lightwave equipment source. The optical power level should be the source level (dBm) less the link loss (dB).

7. If laser light sources are used in the link, test all optical fibers for return loss and ensure that return loss is within laser source specifications.

8. For intermittent errors or poor data throughput, perform a bit error rate test (BERT). The time of errors should be logged to provide clues for cyclic problems. A poor result on the BER test can indicate high optical loss, dirty connectors, defective lightwave equipment, or a poor electrical data connection.

If a fiber break or anomaly is found, determine its location using an OTDR. Follow the OTDR test procedures described in Chap. 14.

If fiber optic test equipment is not available, a quick fiber continuity test can be performed using a flashlight as follows (for multimode fibers only):

1. Turn off and disconnect all fiber lightwave equipment.

2. With the flashlight, shine the light into the fiber to be tested.

3. The light from the flashlight should be visible at the other end of the fiber. The ambient room lighting may need to be dimmed in order to see the fiber light.

4. If the light from the flashlight is not visible, the problem may be that the fiber is broken. Further fiber optic testing using OTDR and power meter is necessary.

Although this flashlight test does not provide any information on the fiber's attenuation, it may help to identify broken fibers or confirm splicing and connection assignments.

CHAPTER 25
FIBER DESIGN FUNDAMENTALS

Designing a fiber optic system can be a complicated process. The designer must consider many factors such as data rate, link attenuation, environment, cable types, fiber types, available equipment, electrical interface types, optical connectors, splicing, protocol, and so on. The complete process is quite involved and therefore beyond the scope of this book.

The process can be simplified when recognized standards or manufacturer's installation instructions are followed. These instructions usually provide sufficient information to select the proper optical fiber type for a simple installation. Other design considerations such as cable types, panels, jumpers, environment, routing, and the like are usually left to the designer to determine.

This chapter will explain how to proceed to select an appropriate fiber for a communication system. Other chapters in this book should be consulted for selecting cable types, equipment, installation locations, etc.

25.1 SINGLE-MODE OR MULTIMODE FIBER

The first decision that must be made is whether to install a single-mode or multimode fiber. Each fiber has its merits.

Advantages of a Single-Mode Fiber Optic System

1. Single-mode optical fiber has the highest possible bandwidth transmission capability and is ideal for long-distance transmission.
2. Single-mode fiber is available for optical wavelengths of 1310 and 1550 nm.

Advantages of a Multimode Fiber Optic System

1. Multimode lightwave equipment is less expensive than similar single-mode equipment because less expensive LEDs (light-emitting diodes) are often used as light sources.

2. Multimode fiber terminating connectors are less expensive than single mode because core alignment tolerances are not as precise.

3. Multimode optical fiber is a standard for LAN communications for distances less than 2 km.

4. Multimode fiber is more tolerant of dust accumulation and connector misalignment.

It can be generally concluded that single-mode fiber installation is best for long-distance (over 2 km) communication systems or high-data-rate transmission (due to the modal dispersion limiting bandwidth of a multimode fiber). Multimode fiber can be deployed for shorter distances as specified by the lightwave equipment manufacturer and applicable standards.

25.2 LINK OPTICAL BUDGET FOR SINGLE-MODE OR MULTIMODE FIBER

Once the fiber type of single-mode or multimode fiber has been decided, all fiber losses in the optical link must be identified and their optical power losses estimated. The losses are summed to obtain the entire fiber link optical power loss. This result can be accomplished using an optical link budget process. The optical link budget is a tabulation of all the losses (or gains) in a fiber optic link. These losses are due to all components in the optical link, such as fiber, connectors, splices, attenuators, WDMs, and so on. It also includes the light source's average output power, receiver sensitivity, and received optical power.

Link Optical Budget: An Example

(a) Optical fiber loss at 1310 nm:	
15.5-km length at 0.35 dB/km	5.4 dB
(b) Splice loss:	
2 splices at 0.1 dB/splice	0.2 dB
(c) Connection loss:	
2 connections at 0.5 dB/connection	1.0 dB
(d) Other component losses	0.0 dB
(e) Design margin	2.0 dB
(f) Total link loss	8.6 dB
(g) Transmitter average output power	−10.0 dBm
(h) Receiver input power:	
(g−f)	−18.6 dBm

(i) Receiver dynamic range	-10 to -25 dBm
(j) Receiver sensitivity at BER 10^{-9}	-25.0 dBm
(k) Remaining margin:	
(h − j)	6.4 dB

Optical fiber loss (item **a** in the optical link budget example) is the total optical fiber loss for the installed fiber cable length at the operating wavelength. This value can be determined from the manufacturer's specifications sheets, or measured. It is specified in the attenuation form dB/km at wavelength and should be multiplied by the fiber length in kilometers to obtain the total fiber loss in dB. The loss can also be measured using a power meter and light source or OTDR.

Splice loss (item **b**) is the total loss to all mechanical or fusion splices in the fiber optic link. During the design stage, this value is normally estimated using manufacturer's specifications. Fusion splices are usually below 0.1 dB, and most mechanical splices are below 0.5 dB loss. Splice loss can also be measured using an OTDR.

Connection loss (item **c**) is the sum of to all connections in the optical fiber link. Losses due to the connectors that are attached on the lightwave equipment can be excluded because they have already been factored into the equipment's specifications by the manufacturer. One connection loss results from two connectors being attached using an adapter. Connection loss can be estimated using the manufacturer's specifications or can be measured with an OTDR or power meter.

Other component losses (item **d**) include the total loss due to other devices in the fiber optic link not included in items **a**, **b**, and **c**, such as splitters or attenuators.

The design margin (item **e**) is an estimated level of loss that will provide sufficient confidence to achieve proper equipment operation during the life span of the system. Over time, all systems degrade to some degree. This should be anticipated in the design stage so that system life can be maximized. Contributing factors include:

- Optical fiber increase in attenuation due to OH ingress
- Decay of the light source output power
- Receiver sensitivity decrease
- Increased link losses due to splice repairs
- Increased connection losses due to contamination

An estimated value of 2 dB is commonly used for this value.

The total link loss (item **f**) is the sum of all optical losses in dB (lines **a** through **e** of the optical link budget).

Transmitter average output power (item **g**) is the average optical output power of the lightwave equipment. This level can be obtained from equipment

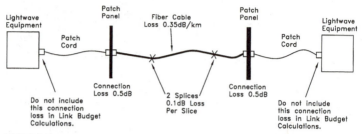

FIGURE 25.1   Link optical budget.

specifications during the design stage but should be measured with a power meter after installation.

Receiver input power (item **h**) is the optical power received by the lightwave equipment. This is calculated by subtracting the total link loss from the transmitter's average output power (line **g** minus line **f** of the optical link budget) or can be measured after equipment has been installed.

Receiver dynamic range (item **i**) is the light-level window in dBm at which a receiver can accept optical power without degrading system performance. The optical receiver is designed to accept light within the manufacturer's predetermined levels (measured in dBm). If the light is too strong, the receiver circuit will saturate, and consequently, the equipment will not operate. This is normally caused by insufficient fiber link attenuation.

To remedy the problem, optical fiber attenuators are inserted in the link until the received optical power is within the receiver's dynamic range. Attenuators can be installed at the equipment connector or at the patch panel connector on the receive end of the fiber. If the attenuators are installed on the transmit fiber, ensure that the return loss measurement is within the equipment's laser specification. Receiver input power (item **h**) should be within the receiver dynamic range (item **i**).

For lightwave equipment designed for short-distance transmissions, the optical receiver may be designed to accept the full light source power (no link attenuation). In this case, fiber attenuators are not required, and this step can be ignored.

Receiver sensitivity at the bit error rate (BER) (item **j**) is the minimum amount of optical power necessary for the lightwave equipment receiver to achieve the specified BER. In an analog system, it is the minimum amount of optical power necessary for the lightwave equipment receiver to achieve the signal-to-noise (S/N) level.

The remaining margin (item **k**) is the additional optical loss that the fiber optic link can tolerate without affecting system performance. It is calculated by subtracting the receiver sensitivity at BER from the receiver input power (line **h** minus line **j** of the optical link budget). This value should always be greater than zero.

If a fiber design uses WDMs, multiple wavelengths, and optical amplifiers, then an optical link budget should be created for each wavelength. In addition, worst-case optical signal-to-noise ratios (OSNRs) should be calculated for each wavelength. For proper system operation, OSNR levels should be within manufacturer's specifications (example: 21 dB OSNR is a standard level for OC48 SONET equipment).

25.3 MULTIMODE FIBER DESIGN

Two important factors to consider in designing multimode fiber optic links are total link loss and maximum link bandwidth. The first factor, total link loss, is the total light power lost in a fiber optic link due to all factors, including connectors, splices, fiber attenuation, cable bends, and the like. Total link loss must be within the lightwave equipment manufacturer's specifications in order for the link to function properly. This is determined by carefully planning an optical link budget, as discussed in the previous section. All factors that contribute or may contribute to optical power loss are included in the link budget.

The second factor, maximum link bandwidth, is the maximum data rate or analog bandwidth that an optical communication system can support with minimal signal distortion. It is limited by lightwave equipment properties and optical fiber parameters. For multimode fiber, fiber optic bandwidth decreases as fiber length increases due to modal dispersion. Therefore, it is important to utilize a cable length that will fit the installation and work with the given lightwave equipment.

LAN Installations.   For LAN systems, appropriate industry standards (including the TIA/EIA 568A cabling standard) should be consulted to determine proper fiber types and cable lengths. These industry standards are recognized by many LAN equipment manufacturers whose equipment is designed to function properly within the standards (see Chap. 19 for more information). Global Engineering is a good source for many of the standards.

Non-LAN Installations.   For non-LAN installations, such as control circuitry in industrial sites, the equipment manufacturer may not have designed the equipment to work within the TIA/EIA standard. Therefore, a fiber design should be performed in order to determine proper fiber type and length.

Lightwave equipment manufacturers will usually recommend an optical fiber type or a number of different optical fiber types that can be used successfully with their equipment. These optical fiber types have been tested with their equipment in a standard point-to-point configuration for the indicated fiber's maximum length and maximum loss. Equipment will operate successfully if the recommended fiber types are implemented within the fiber loss and length restrictions. These restrictions may be listed by the equipment manufacturer in tabular form, as shown in the following example:

Fiber type (μm)	Fiber attenuation (dB/km)	Fiber NA	Fiber bandwidth (Mhz × km)	Maximum loss (dB) at 850 nm	Maximum length (km)
50/125	3.0	0.20	50	2.0	0.6
50/125	2.7	0.20	50	2.0	0.7
62.5/125	3.5	0.29	50	5.0	1.4
62.5/125	3.0	0.29	50	5.0	1.6
100/140	5.0	0.29	50	9.5	1.5
100/140	4.0	0.29	50	9.5	1.8

The first three columns of this table list optical fiber specifications for a number of commercially available optical fibers.

The fiber bandwidth value is the fiber's normalized bandwidth (to 1 km) and is the lowest optical fiber bandwidth that the lightwave equipment manufacturer recommends for the installation. Optical fibers with larger bandwidths are acceptable for the installation (such as 100 Mhz × km in this example).

The maximum loss and maximum length should not be exceeded for the selected fiber type. The maximum loss should always be greater than, or equal to, the total link loss.

The maximum length is the total optical fiber length between the terminating lightwave equipment. The length represents the limit for loss due to fiber attenuation and for bandwidth due to fiber dispersion. It should not be exceeded even if the total calculated optical link loss is below the maximum loss.

As shown in this table, cable length increases with the size of the core diameter. This is due to the increase of the light power coupling from an LED light source into a fiber as a result of the fiber's larger-diameter core and greater numerical aperture (assuming dispersion is not a factor).

If the user has a choice, he or she should select a standard fiber core size for the installation. The 62.5/125-μm fiber has been the standard fiber for many years; however, now the 50/125-μm fiber is returning in popularity due to its larger bandwidth.

Method

1. Obtain the following information from the lightwave equipment manufacturer:

   - Optical fiber diameter recommendations—50/125, 62.5/125, or 100/140
   - Maximum optical fiber attenuation (dB/km) recommendation
   - Optical fiber numerical aperture (NA) recommendation
   - Maximum fiber bandwidth (Mhz × km) at operating wavelength recommendation
   - Optical fiber maximum length recommendation

- Equipment's maximum loss specification; if the maximum loss is not specified, it can be calculated from the receiver's sensitivity and transmitter's average output power specifications:

Maximum loss = transmitter's average output power − receiver's sensitivity

2. From the fiber optic installation plan determine:
   - Total fiber optic cable length
   - Number of required optical splices and the loss per splice
   - Number of fiber connections and the loss per connection
   - Design margin
   - Optical losses due to any other components in the system
3. Complete the optical budget as described in Sec. 25.2.
   - Optical fiber loss at operating wavelength: kilometer length at dB/km
   - Splice loss: splices at dB/splice
   - Connection loss: connections at dB/connection
   - Other component losses
   - Design margin
   - Total link loss
   - Transmitter's average output power
   - Receiver's input power
   - Receiver's dynamic range
   - Receiver's sensitivity
   - Remaining margin
4. The remaining margin should be greater than zero for proper system design. If it is not, reexamine all loss values to reduce total link loss.

Selecting Optical Fiber Example. A fiber optic link is to be designed to provide point-to-point data communications between two computers (see Fig. 25.2). Lightwave communication equipment that is compatible with the computer equipment (proper electrical interface and communication protocol) has been selected. The equipment manufacturer's recommended optical fiber specification is presented as a table later in this example. In addition, the receiver's dynamic range is from zero to full transmitter output.

The cable length through an outdoor route has been measured and is 1.2 km. Due to the nature of the installation, four separate fiber cables will be required to complete the link. A patch panel is requested at both ends for easy connection of patch cords. What type of optical fiber should be selected for the installation? (See the following table.)

The two major considerations for optical fiber selection are total link loss and optical fiber length. The total length between lightwave terminating equipment has been measured at 1.2 km. Any future extension to the fiber link should be considered here. Both optical link loss and fiber cable length will increase if an extension fiber is added to accommodate additional future requirements. If not properly planned, an extension to the fiber cable may not be possible. For this example, future extension of the link is not required.

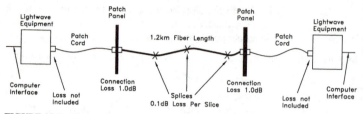

FIGURE 25.2 Fiber optic computer link example.

The first step is to compile all known information to determine an optical link budget for the installation.

The fiber optic cable length is 1.2 km. Three splices are required to connect the four sections of cable. Because the cables will be spliced outdoors, the fusion splicing method has been selected, with a maximum loss of 0.1 dB per splice. Two patch panels are to be used with patch cords connecting the equipment. The fiber optic cable is to be directly terminated with connectors. Connection loss at each patch panel is 1.0 dB. Patch cord connection loss at equipment is not added to the link budget because it is already factored into the manufacturer's specifications. Patch cords are quite short at 3 m, so their fiber attenuation is minimal and is therefore ignored.

Fiber size (μm)	Fiber attenuation (dB/km)	Fiber NA	Fiber bandwidth (Mhz × km)	Maximum loss (dB) at 850 nm	Maximum length (km)
50/125	3.0	0.20	50	2.0	0.6
50/125	2.7	0.20	50	2.0	0.7
62.5/125	3.5	0.29	50	5.0	1.4
62.5/125	3.0	0.29	50	5.0	1.6
100/140	5.0	0.29	50	9.5	1.5
100/140	4.0	0.29	50	9.5	1.8

1. The following information has been obtained from the lightwave equipment and cable manufacturers:

- Recommended optical fiber type and diameter(s), in table form for six optical fibers
- Recommended optical fiber maximum attenuation (dB/km) at operating wavelength, in table form for six optical fibers
- Recommended optical fiber numerical aperture (NA), in table form for six optical fibers
- Recommended optical fiber bandwidth (Mhz × km) at operating wavelength, in table form for six optical fibers
- Optical fiber maximum length, in table form for six optical fibers

- Equipment maximum loss specification for optical fiber type used, in table form for six optical fibers
- Equipment receiver sensitivity at BER not provided
- Equipment transmitter average output power not provided
- Equipment receiver dynamic range, from full source power to minimum receiver sensitivity (a full dynamic range)

2. From the fiber optic installation plan:

- Total fiber optic link length is 1.2 km.
- Number of required optical splices is three at 0.1 dB per splice.
- Number of fiber connections is two at 1 dB per connection.
- Design margin estimate is 2 dB.
- Optical losses due to any other components in the system are zero.

   From the lightwave equipment manufacturer's recommended optical fiber list, the distance criteria can be met with any of the 62.5/125 or 100/140 optical fibers. The first choice for an optical fiber is a standard 62.5/125 with 3.0-dB/km loss. This is used for a preliminary optical budget calculation as follows.

3. Optical budget:

Optical fiber loss at 850 nm:	
1.2-km length at 3.0 dB/km	3.6 dB
Splice loss:	
Three splices at 0.1 dB/splice	0.3 dB
Connection loss:	
Two connections at 1.0 dB/connection	2.0 dB
Other components' loss	0
Optical margin	2.0 dB
Total link loss	7.9 dB

   Therefore, using the 62.5/125-μm, 3.0-dB/km optical fiber will result in a total fiber link loss of 7.9 dB. This is higher than the lightwave equipment manufacturer's maximum loss of 5 dB and therefore cannot be used.

A second choice will be the 100/140 fiber with 4-dB/km attenuation. The link budget then looks as follows:

Optical fiber loss at 850 nm:	
1.2-km length at 4.0 dB/km	4.8 dB
Splice loss:	
Three splices at 0.1 dB/ splice	0.3 dB
Connection loss:	
Two connections at 1.0 dB/connection	2.0 dB
Other components' loss	0
Optical margin	2.0 dB
Total link loss	9.1 dB

The total link loss using the 100/140-μm at 4 dB/km fiber is 9.1 dB. This is less than the lightwave equipment manufacturer's maximum loss of 9.5 dB for this fiber type. Therefore, the maximum loss criterion has been satisfied. The lightwave equipment manufacturer's maximum length for this fiber type is 1.8 km, above the required 1.2-km installation length. This length, therefore, satisfies the length criterion. Consequently, this optical fiber type can be used for the installation.

Many lightwave equipment manufacturers will recommend that one standard optical fiber (62.5/125) be used with their equipment. This simplifies the procedure.

The following example shows calculations based on this information.

Standard Optical Fiber Extension Used for a LAN System Example. A LAN segment in a plant is to be extended using fiber optics. Two LAN optical fiber modems (media converters) designed specifically for this purpose are being considered for the link. The modem manufacturer provides the following information for the optical fiber to be used with the equipment:

Equipment operating wavelength	850 nm
Optical fiber type	Multimode 62.5/125, NA = 0.29
Optical fiber bandwidth	100 MHz $\times$ km
Maximum fiber attenuation	5.0 dB/km
Maximum optical fiber length	1 km
Receiver dynamic range	Full range

The modem manufacturer specifications assume that the optical fiber will not be spliced, that only two connectors at the equipment will be required, and that no other attenuation will be added to the link.

The length of the fiber optic link is measured to be 0.7 km. The fiber optic cable will be dedicated and will not require patch panels or splices. What optical fiber should be purchased for the link to operate? What will the design look like?

This design is simple, and the equipment manufacturer provides all the information required for proper optical fiber purchase.

The measured optical fiber length will be 0.7 km, which is below the manufacturer's maximum length criteria for a 100-MHz-$\times$-km fiber. No splices or connectors (other than directly terminating cable connectors) are used. The only loss will result from optical fiber attenuation. The fiber to select should provide the same or better specification than the manufacturer's recommendation. The design will resemble that which is shown in Fig. 25.3, a dedicated fiber optic cable with no intermediate splice or connections. The fiber should be multimode 62.5/125 μm, NA 0.29, bandwidth of 100 MHz $\times$ km or better, and loss of 5 dB/km or less.

If splices or connectors are to be added, then a link budget should be determined. For example, if two connectors and one splice are added to the fiber link, then the budget will be as follows:

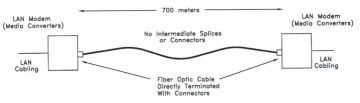

FIGURE 25.3 Fiber optic LAN extension.

Optical fiber loss at 850 nm:	
0.7 km length at ? dB/km	? dB
Splice loss:	
One splice at 0.1 dB/splice	0.1 dB
Connection loss:	
Two connections at 1.0 dB/connection	2.0 dB
Other components' loss	0 dB
Optical margin	2.0 dB
Total link loss	5.0 dB

The modem manufacturer's maximum fiber attenuation is converted to a maximum loss by multiplying by the 1-km maximum length. The resultant 5 dB is used as a total link loss. Rearranging the equation, optical fiber loss is calculated to be 0.9 dB, and the optical fiber attenuation should be 1.2 dB/km or less:

$$\text{Optical fiber loss} = 5.0 - 2.0 - 2.0 - 0.1$$

$$= 0.9 \text{ dB}$$

$$\text{Optical fiber attenuation} = 0.9 \text{ dB}/0.7 \text{ km}$$

$$\text{Maximum optical fiber attenuation} = 1.2 \text{ dB/km}$$

A maximum attenuation of 1.2-dB/km optical fiber is required if two connectors and a splice are to be added to the link. The fiber must have a 100-MHz-$\times$-km fiber bandwidth or better.

In the next example, the equipment manufacturer provides nominal optical output power and receiver sensitivity instead of attenuation or loss level. This requires additional calculations but is not complicated.

Nominal Optical Output Power and Receiver Sensitivity Example. A fiber optic link is to be used to connect a remote video surveillance camera to a monitor that is 3 km away (see Fig. 25.4). The camera is a high-quality model with a 10-MHz bandwidth and standard NTSC video output. A manufacturer of lightwave equipment that can convert the 10-MHz NTSC electrical signal to an optical transmission is found, and it has the following equipment specifications:

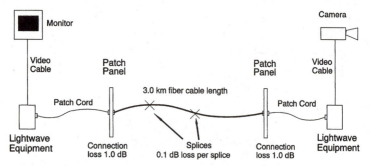

FIGURE 25.4 Fiber optic video link example.

Equipment operating wavelength	1310 nm
Optical fiber type	multimode 62.5/125, NA = 0.29
Optical fiber bandwidth	300 MHz × km
Nominal optical output power	−15 dBm
Receiver sensitivity	−25 dBm at S/N 68 dB
Maximum optical fiber length	3 km
Receiver dynamic range	−20 to −25 dBm

Two patch panels and two splices will be required for the installation. Can this lightwave equipment be used for this installation? If so, what optical fiber is required?

The first step is to determine an optical link budget for the installation. The fiber optic cable length is 3 km. Two splices are required to connect the three sections of cable. Because the cables will be spliced outdoors, the fusion splicing method has been selected with a loss of 0.1. The cable will be directly terminated with connectors. Two patch panels are to be used with patch cords to connect the equipment. Connection loss at each patch panel is 1.0 dB. Patch cord connection loss at the equipment is not added to the link budget because it is already factored into the manufacturer's specifications. Patch cords are quite short at 3 m, so their fiber attenuation is minimal and is ignored.

1. The following information was obtained from the equipment manufacturer:

- Optical fiber diameter recommendation 62.5/125
- Optical fiber attenuation to be determined
- Optical fiber numerical aperture 0.29
- Optical fiber bandwidth at operating wavelength 300 MHz × km
- Optical fiber maximum length 3 km
- Equipment maximum loss not provided
- Equipment receiver sensitivity −25 dBm at 68 S/N

- Equipment transmitter average output power -15 dBm
- Equipment receiver dynamic range is -20 to -25 dBm

2. From the fiber optic installation plan:

- Total fiber optic link length is 3 km
- Number of required optical splices is two at 0.1 dB per splice
- Number of fiber connections is two at 1 dB per connection
- Design margin estimate 2 dB
- Optical loss due to any other components in the system is zero

3. The optical budget calculations is as follows:

(a)	Optical fiber loss at 1310 nm: 3.0-km length at ? dB/km	? dB
(b)	Splice loss: Two splices at 0.1 dB/splice	0.2 dB
(c)	Connection loss: Two connections at 1.0 dB/connection	2.0 dB
(d)	Other components' loss	0 dB
(e)	Optical margin	<u>2.0 dB</u>
(f)	Total link loss	? dB
(g)	Transmitter's average output power	-15.0 dBm
(h)	Receiver's input power $(g-2 f)$	? dBm
(i)	Receiver's dynamic range	-20 to -25 dBm
(j)	Receiver's sensitivity at 68 dB S/N	-25 dBm
(k)	Remaining margin $(h-j)$	0 dB

Working backwards, we can solve for receiver input power using a remaining margin of 0 dB and receiver sensitivity of -25 dBm:

$$\text{Receiver input power} = \text{Remaining margin} + \text{receiver sensitivity}$$

$$= 0 \text{ dB} + (-25 \text{ dBm})$$

$$= -25 \text{ dBm}$$

Next we solve for total link loss:

$$\text{Total link loss} = \text{Transmitter average output power} - \text{receiver input power}$$

$$= -15 \text{ dBm} - (-25 \text{ dBm})$$

$$= 10 \text{ dB}$$

To determine the optical fiber loss use the following formula:

$$\text{Optical fiber loss} = \text{total link loss} - \text{connection loss}$$
$$- \text{optical margin} - \text{splice loss}$$

$$\text{Optical fiber loss} = 10.0 - 2.0 - 2.0 - 0.2$$

$$= 5.8 \text{ dB}$$

The optical fiber loss is divided by the total cable length to determine the optical fiber attenuation:

$$\text{Optical fiber attenuation} = 5.8 \text{ dB/3.0 km}$$

$$= 1.9 \text{ dB/km}$$

The 62.5/125 optical fiber (NA = 0.29) to be used for this installation should have an attenuation no greater than 1.9 dB/km at 1310 nm. To satisfy the bandwidth criterion, the cable should have a bandwidth of 300 MHz × km and be 3 km or shorter in length.

In these examples, a 2-dB optical design margin was used to account for any future contingencies such as additional fiber splices, dirty connectors, or cable extension.

25.4 SINGLE-MODE FIBER DESIGN

For single-mode fiber design, total link loss and bandwidth limiting factors must be considered. Total link loss is the total light power lost in a fiber optic link due to all factors, including connectors, splices, fiber attenuation, cable bends, WDMs, and any other devices connected to the fiber. Total link loss must be within the lightwave equipment manufacturer's specifications in order for the equipment to function properly. A detailed optical link budget, as discussed in the prior section, should be planned for the entire fiber optic system. All factors that contribute or may contribute to optical power loss are included in the link budget.

The bandwidth limiting factor is fiber dispersion (also PMD for 2.5-Gbps or higher systems). Dispersion introduces a path loss penalty and limits bandwidth if not compensated.

Basic Method

1. Obtain the following information from the lightwave equipment manufacturer:

   - Maximum allowable fiber optical loss in dB
   - Maximum dispersion in ps/nm and dispersion penalty in dB
   - Maximum polarization mode dispersion or differential group delay (for OC-48 and OC-192 systems)

   If the maximum optical loss is not specified, it can be calculated from the receiver's sensitivity and transmitter's average output power specifications:

   Maximum loss = transmitter's output power − receiver's sensitivity

2. From the fiber optic installation plan, determine:

   - Total fiber optic cable length
   - Number of required optical splices and the loss per splice
   - Number of fiber connections and the loss per connection
   - Design margin
   - Optical losses due to any other components in the system

3. Complete an optical link budget as described in previous section:

   - Optical fiber loss at operating wavelength: kilometer length at dB/km
   - Splice loss: splices at dB/splice
   - Connection loss: connections at dB/connection
   - Other component losses
   - Design margin
   - Total link loss
   - Transmitter's average output power
   - Receiver's input power
   - Receiver's dynamic range
   - Receiver's sensitivity
   - Remaining margin

4. The remaining optical margin should be greater than zero for proper system design. If it is not, reexamine all loss values to reduce total link loss.

5. Determine total fiber link chromatic dispersion by measurement or calculation and compare to manufacturer's specification. If dispersion is greater than manufacturer's specification, then appropriate dispersion compensation must be added.

6. For OC-48 and OC-192 systems, polarization mode dispersion (PMD) should be measured and compared to manufacturer's specifications.

7. Fiber optic link total return loss should be measured for each fiber. Values are then compared to manufacturer's specifications. If greater than specification, the individual component's reflectance should be reduced along the link.

OC-48 System Design Example. A SONET OC-48 system is to be installed to provide communications between two cities. The fiber optic cable has already been installed and has a length of 73 km. A power meter test for fiber loss resulted in measurements of 18.3 dB loss at 1550 nm and 26.5 dB loss at 1310 nm.
    SONET equipment specifications are:

Equipment operating wavelength	1310 or 1550 nm
Optical fiber type	single mode
Nominal optical output power	0 dBm
Receiver sensitivity	−27 dBm
Receiver overload	−15 dBm
Transmitter maximum return loss	−25 dB
Maximum dispersion	1500 ps/nm @ 1550 nm, penalty −2.5 dB
DGD penalty due to PMD	8 to 10 ps, penalty 0.5 dB

Two patch panels with jumpers will be used at both ends. Can this light-wave equipment work for this fiber installation?

First let's check to see if the system will be within the optical budget at 1310 nm. Lasers at this wavelength are generally less expensive than the 1550-nm lasers, and therefore, if they can be used, system cost can be kept to a minimum. Our optical budget will be as follows:

Optical fiber loss at 1310 nm:	
73-km length	26.5 dB
Splice loss:	
Two splices at 0.1 dB/splice	included
Connection loss:	
Two connections at 0.3 dB/connection	0.6 dB
Other losses (DGD)	0.5 dB
Optical margin	2.0 dB
Total link loss	29.6 dB
Transmitter's average output power	0 dBm
Receiver's input power	−29.6 dBm
Receiver's dynamic range	−15 to −27 dBm
Receiver's sensitivity	−27 dBm
Remaining margin	n/a

This calculation results in receiver optical power of −29.6 dBm, which is much less than the −27 dBm specified for the receiver. The system will not work using 1310-nm lasers.

Next we try 1550-nm lasers:

Optical fiber loss at 1550 nm:	
73-km length	18.3 dB
Splice loss:	
Two splices at 0.1 dB/splice	included
Connection loss:	
Two connections at 0.3 dB/connection	0.6 dB
Other losses (DGD & Disp.)	3.0 dB
Optical margin	2.0 dB
Total link loss	23.9 dB
Transmitter's average output power	0 dBm
Receiver's input power	−23.9 dBm
Receiver's dynamic range	−15 to −27 dBm
Receiver's sensitivity	−27 dBm
Remaining margin	3.1 dB

This result is within the receiver's optical range; therefore, the optical budget is within specification.

Next, the link dispersion should be measured. The field measurement of the chromatic dispersion has resulted in a value of 1350 ps/nm. The equipment can tolerate a maximum dispersion of 1500 ps/nm, which suggests the dispersion is within specification. However, the manufacturer also specifies a path penalty of 2.5 dB, which was already added to our budget estimate.

Polarization mode dispersion should be measured for OC-48 and OC-192 long-haul transmission, especially if optical amplifiers are to be deployed. A field measurement results in a DGD value of 8 ps. Equipment specifications call for a 0.5-dB path penalty for an 8-ps delay. This penalty was also added to our link budget estimate for a total link loss of 23.9 dB.

One last check for return loss should be performed. These results can only be accurately determined by field measurement. Measurement of the fiber's return loss resulted in a worst-case value of −35 dB, which is within the equipment specification of −35 dB. Therefore, this fiber link can be used for this OC-48 system.

25.5 FIBER NETWORK TOPOLOGIES

Fiber optic networks should be designed to provide a flexible and versatile system that will allow the full benefits of the optical fiber to be realized. The traditional practice of system installation is to design and dedicate a fiber cable to each new application. Not only can this be costly, but it can also greatly limit future cable potential. Systems should be installed to a carefully designed, fiber optic cable deployment network plan. Then, for example, when a single point-to-point application is installed, it can be upgraded to a future planned network.

Network topologies can be classified as logical or physical. Logical topology describes the method by which the different nodes of the network communicate with each other. Physical topology is the actual physical layout of the cabling and nodes in the network (see Fig. 25.5).

There are four standard logical topologies that optical fiber can support:

1. Point-to-point

2. Star

3. Bus

4. Ring

The following sections briefly describe each of these topologies.

1. **Logical point-to-point.** Logical point-to-point topology links are two devices that directly communicate with each other (see Fig. 25.6). An example would be dedicated data links such as T1, T3, linear SONET, and optical modem links.

2. **Logical star.** The logical star topology is an arrangement of point-to-point links that all have a common node (see Fig. 25.6). Applications include telephone PBX and multistation video monitoring systems.

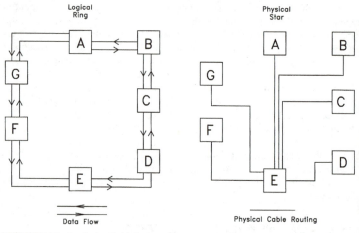

FIGURE 25.5   Logical and physical topologies.

3. **Logical bus.** In logical bus topology, all devices are connected to one transmission bus, usually with coaxial cable (see Fig. 25.7). On this bus, transmission occurs in both directions. When one device transmits information, all the other devices receive the information simultaneously. This bus topology is the LAN IEEE 802.3 and 802.4 standard. Applications include Ethernet and token bus.

4. **Logical ring.** Logical ring topology has all nodes connected into a ring (see Fig. 25.7). Transmission occurs in one or both directions in the ring.

25.5.1   Physical Topologies

The optical fiber physical topology can be implemented in the same configuration as the logical topology, except for the bus topology:

1. Point-to-point
2. Star
3. Ring

1. **Physical point-to-point.** This topology is common where using fiber modems (media converters) or in SONET linear systems. Both logical bus and ring systems can be initially deployed as a point-to-point topology, and later converted to physical ring or bus as the network grows and proper cables are in place to facilitate the topology.

   Advantages of a point-to-point topology:

   • Easiest to install and administer
   • Easiest to troubleshoot

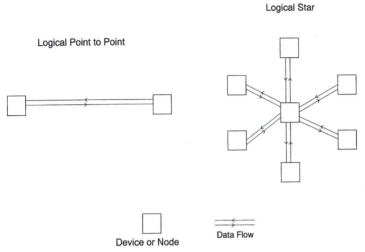

FIGURE 25.6   Logical point-to-point and star topologies.

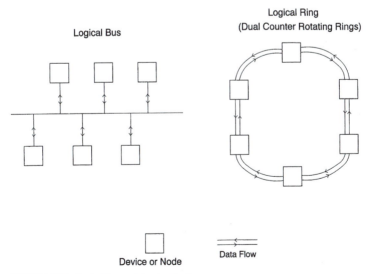

FIGURE 25.7   Logical bus and ring topologies.

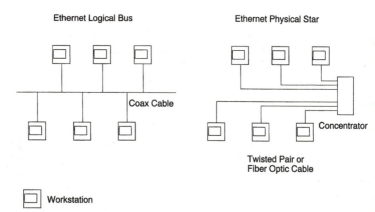

FIGURE 25.8   Logical bus topology as a physical star.

Disadvantages:

- Does not provide any kind of diverse routing protection of traffic. If the cable between the two points is cut, then all traffic is lost until the cable is restored.
- Can be difficult to administer for large networks, if point-to-point systems are not networked together.

2. **Physical star.** This topology is common for LAN networks. It has a centralized hub which allows for convenient centralized access to transmissions to all nodes. Ethernet systems are commonly configured in this topology even when they are logical bus systems.

Advantages of a star topology:

- Flexibility and ability to support many applications and all other logical topologies, such as bus (see Fig. 25.8) and ring (see Fig. 25.9).
- A centralized optical fiber cross-connect location allowing easy fiber maintenance and administration.
- Recommended in EIA 568 standard for commercial building wiring.
- Many existing duct and conduit routes are often a star configuration.
- Easier system expansion.

Disadvantages:

- A cable cut will result in the connected device or node to fail.
- It requires more fiber optic cable length to implement than a physical ring topology.

3. **Physical ring.** Ring technology is the best for protection of traffic. Most ring systems are designed to recover from a cable cut without losing traffic. If a cable cut or failure occurs anywhere in the ring, traffic will be rerouted around the other end of the ring to its destination (see Fig. 25.10), increasing network reliability. Common applications of ring topologies are: SONET rings, FDDI, and token ring LAN (see Chap. 18, SONET/SDH, for more details).

Advantages of a ring:

- Ring system can survive a cable failure or cut by rerouting traffic around the other end of the ring to its destination.
- Uses less cable than similar star system.

Disadvantages:

- More difficult to install, maintain, and repair.

Logical Ring & Physical Ring

Logical Ring & Physical Star

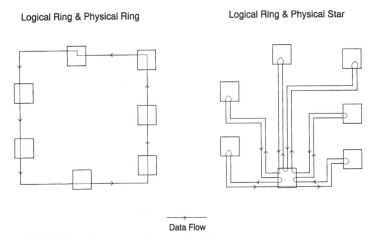

Data Flow

FIGURE 25.9   Logical ring topology as a physical star.

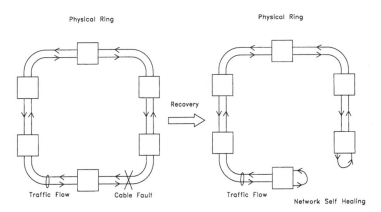

FIGURE 25.10   Traffic rerouting in a physical ring topology.

CHAPTER 26
PERSONNEL

An aspect often overlooked in many system implementations is the need for properly trained personnel to install, operate, and maintain a fiber optic system. On the exterior, fiber optic cable may look similar to electrical cable; special handling and installation techniques, however, should be used. Many people fall into the trap of trying to "throw a fiber optic link into commission" with little or no knowledge of the subject. Although this approach may work on occasion, many problems can be encountered during and after the installation, and the full potential of the fiber cable may not be realized.

The best way to ensure that the system is designed and installed properly is to employ personnel who are well trained in this field. Many companies provide this expertise or offer training to employees. The best instruction combines classroom fiber optic theory with practical, hands-on fiber optic system training.

The hands-on approach should include: testing techniques that use a light source and power meter; fiber optic cable stripping, splicing, and connectorization; and optical fiber cleaning—all combined with proper safety instruction.

The instruction should remain practical (unless a design course is sought) and include such topics as: light transmission in optical fibers, fiber optic components, cable types and applications, testing equipment and procedures, and cable handling, installation, and maintenance and repair techniques, with safety concerns always emphasized.

Training is also offered on an "in-house" basis. This allows employees to learn at the job site and become familiar with fiber optics in their work environment.

A good introductory practical course can be up to four full days in length, with additional days required to accommodate the hands-on training of each student. It is also recommended that a knowledgeable supervisor be at the job site to ensure that a first-time fiber optic installation is done properly. Mistakes can easily occur and can be quite expensive to remedy.

The following is an example of a typical agenda for a basic fiber optic course:

Day 1
Introduction to Optical Fiber Communication

- History of optical fiber
- Advantages and drawbacks of optical fiber communication
- Practical examples of optical fiber communication
- Data, voice, and video communication over fiber

Introduction to Optical Fiber

- Optical fiber light spectrum
- Transmission techniques
- Analog signals and digital signals

Optical Fiber Composition

- Fiber diameters
- Light transmission in a fiber
- Multimode fiber
- Step index fiber
- Graded index fiber
- Single-mode fiber
- Optical power loss
- Data rate limitations
- Optical fiber example

Fiber Optic Safety

- Safety precautions when working with optical fiber

Fiber Optic Cable Composition

- Loose-tube and tight-buffered cable
- Cable jackets
- Strength members
- Fiber counts and cable size
- Cable specifications
- Cable procurement

Day 2
Method of Handling Fiber Optic Cable

- Bending radius
- Pulling tension
- General care

Outdoor Fiber Optic Cable Installation

- Duct/innerduct installation
- Lubrication
- Pulling tape
- Buried cable installation
- Cable pulling technique
- Aerial installation

Indoor Fiber Optic Cable Installation

- Conduit placement
- Fire code rating
- Riser installation
- Equipment locations

Splicing and Termination Techniques

- Splice enclosures and splice trays
- Patch panel termination
- Fusion splicing and mechanical splicing
- Connector installation

Day 3–4
Hands On

- Fiber testing
- OTDR testing
- Power meter and source tests
- Splicing—mechanical and fusion
- Fiber termination with connectors
- Proper cleaning of fiber connectors
- Fiber scope operation
- Proper handling of fiber cable
- Fiber stripping and cleaning techniques

CHAPTER 27
DARK FIBER LEASING

Leasing dark fiber has become common practice in the telecom industry. The term "dark fiber" refers to fibers that are not connected to any lightwave equipment. A company that owns a fiber cable facility (lessor) leases individual dark fibers to others (lessees) on a monthly or yearly basis. It is up to the lessee to add proper lightwave equipment to the dark fibers in order to establish a communications link.

When negotiating a dark fiber lease, the following points should be considered.

Fiber Access Location.   It is important to clearly identify the locations at which to access both ends of the fiber cable, including building address, rack position, fiber patch panel position, and fiber patch panel number(s). At these locations, fiber access agreements may need to be negotiated. Space and power agreements also may need to be negotiated for the installation of lessee terminating lightwave equipment.

Fiber Termination.   Fiber connector types that will terminate the leased fibers should be specified. For some fiber leases, only bare fiber (no connector) is provided, and it is up to the lessee to splice pigtails, or extension fiber, to the leased fiber.

Fiber Count.   The number of fibers that will be available to the lessee throughout the entire cable route should be verified.

Fiber Windows.   Some companies may only lease fiber windows or DWDM wavelengths in a fiber, and not the complete fiber. For example, if the fiber's 1310-nm window is already being used, then the lessor may only allow the 1550-nm window to be leased. If this is the case, the available window wavelength range or DWDM wavelengths should be specified to the lessee. If the entire fiber is available to the lessee, this should be stated in the lease, and the wavelength range that is available to the lessee should be stated.

Fiber Cable Route.   A map of the fiber cable route would be useful to the lessee, although this information may be difficult to obtain from the fiber cable provider. Fiber cable route information is often a closely guarded company secret. If the leased fiber cable route is known, diverse cable routes can be planned for future expansion.

Fiber Type.   Fiber type should be specified for the entire fiber cable length. Single-mode fiber types can include NDSF, DSF, NZDSF, etc. The fiber type should be compatible with the lightwave equipment engineering design. Some fiber types may not be compatible with certain lightwave systems (for example, DWDM systems). The design engineer should be consulted for proper fiber types.

Fiber Loss.   End-to-end maximum fiber optical loss, in dB at a specified wavelength, should be guaranteed for the entire length of the lease by the lessor. If a number of windows or wavelengths are to be leased, then maximum loss in each window or wavelength should be specified. This point is important to consider ensuring that the lessor properly maintains the fiber and does not add any unnecessary losses into the fiber link during fiber repair or maintenance.

Return Loss.   Maximum reflected optical power (return loss) should be guaranteed by the lessor. Too much reflected optical power can cause problems with laser sources.

Restoration Time.   In the event of a fiber cable cut or other fiber outage, the lessor should guarantee a maximum time to fiber restoration. Time to restore a fiber can vary greatly, depending on factors such as lessor commitment, availability of splicing crews, and travel time to splicing location. The lessee should be given a 24/7 (staffed 24 hours a day, 7 days a week) emergency contact number or numbers to report a fiber outage or other fiber problems.

   Also, the lessor's maximum fiber down time in minutes per year, or fiber availability, should be made available to the lessee.

Fiber Test Results.   OTDR and power meter test results for each fiber, and each window or wavelength leased, should be provided as part of the acceptance package of a leased fiber facility. Additional useful test results include return loss, dispersion, and PMD. Test results provided by a third independent party can help to ensure unbiased results.

Length of Lease.   The length of the lease, and payment terms should be clearly stated.

   The IRU (Indefeasible Rights of Use) agreement is often used when leasing dark fiber. The IRU is an irrevocable contractual agreement for the right to use dark fiber bandwidth, over a certain time, as stipulated in the agreement.

APPENDIX A

GLOSSARY OF TERMS AND ACRONYMS

802.3   A local area network protocol that uses CSMA/CD for medium access control and a bus topology defined in layers 1 and 2 of the OSI protocol stack.

802.4   A local area network protocol known as token bus uses a token-passing access method and a bus topology. Originated by General Motors and targeted for the manufacturing environment. It is defined in layers 1 and 2 of the OSI protocol stack.

802.5   A local area network protocol known as token ring uses a token-passing with priority and reservation access method and a star wired ring topology. Originated by IBM. It is defined in layers 1 and 2 of the OSI protocol stack.

absorption   The loss of optical power in fiber, resulting from the conversion of light to heat. Caused by impurities, OH migration, defects, or absorption bands.

acceptance angle   The angle at which all incident light is totally and internally reflected by the optical fiber core. Acceptance angle = sin NA. Also known as maximum coupling angle.

adapter   A mechanical device used to align and join two fiber optic connectors. It is often referred to as a coupling, bulkhead, or interconnecting sleeve.

ADM   Add/drop multiplexer.

ADSS   All dialectic self-supporting cable.

analog   A waveform format that is continuous and smooth and is used to indicate infinite levels of signal amplitude. *Also see* digital.

angstrom   Unit of length equal to 10^{-10} meter or 0.1 nanometer.

ANSI   American National Standards Institute.

APD   Avalanche photo diode. A diode used for detecting very small quantities of light.

aramid yarn   A light material, usually yellow or orange, that provides strength and support to fiber bundles in a cable. Kevlar is a particular type of aramid yarn that has very high strength.

armor   Additional protection between cable jacket layers usually made of corrugated steel.

asynchronous   A signal that is not synchronized to a network clock.

ATM   Asynchronous transfer mode. A communication protocol standard using 53-byte packets defined in layer 2 (data link) of the OSI protocol stack.

attenuation, optical fiber   The diminution of light in the optical fiber. Generally expressed without a negative sign in dB or dB/km. When specifying attenuation, it is important to note the applicable wavelength. Attenuation of an optical fiber is different for different optical wavelengths.

attenuator, optical   A device that reduces the fiber optic light beam intensity. Usually inserted at a connection point.

backbone cabling   The portion of telecommunication cabling that connects telecommunication closets, equipment rooms, buildings, or cities. It is a transmission medium (usually fiber optics) that provides a high-speed connection to numerous distributed facilities.

bare fiber adapter   A bare fiber adapter is an optical fiber connector designed to temporarily connect an unterminated optical fiber to a connector. This allows for quick testing of unterminated fibers.

baud rate   The number of electrical transitions of a transmitted digital signal, per second. Not the same as data rate. A modem can have a transmission data rate higher than its baud rate.

bidirectional coupler (WDM)   A WDM that allows optical transmissions in the common fiber in both directions.

binary *n*-zero suppression   A line coding system for digitally transmitted signals. The *n* represents the number of zeros that are replaced with a special code to maintain pulse density required for synchronization. Typically *n* is 3, 6, or 8.

bit   One binary digit.

bit error rate (BER)   The ratio of bits received in error to the number of bits sent. A BER (bit error rate) of 10^{-9} is common (a billion bits sent with one bit received in error).

bit error rate tester (BERT)   An instrument that measures the amount of bits transmitted incorrectly by a digital communication system.

BLSR   Bilateral line switched ring.

bps   In the telecommunication world, this abbreviation means bits per second. It is usually preceded with a prefix k for kilo meaning thousand, M for mega meaning million, or G for giga meaning billion. Note that in the computer world this abbreviation can also mean bytes per second. It is sometimes, but not always, written with a capital B, as Bps, to indicate bytes. Before interpreting the meaning of this abbreviation, the reader should identify the context in which it is presented.

bridge   Connects two or more similar LANS in layer 2 of the OSI protocol stack.

broadband   Data rates at or greater than 45 Mbps (DS3).

brouter   A vendor device that acts both as a bridge and router.

buffer   A protective cover of plastic or other material, usually color coded, covering the optical fiber. A buffer can either be tight, as in a tight-buffered cable and adhered directly to the optical fiber coating, or it can be loose, as in a loose-tube cable, where one or more fibers lie in the buffer tube loosely. The buffer must be stripped off for cleaving and splicing.

byte   Eight adjacent binary digits (bits).

CAD   Computer-aided design.

CAM   Computer-aided manufacturing.

CATV   Community antenna television, also known as cable television system.

CCITT (now ITU-T)   Consultative Committee on International Telephone & Telegraph. An international committee that develops and recommends standards for telecommunications.

CCTV   Closed-circuit television.

CDDI   Copper distributed data interface. A protocol standard similar to FDDI but using an unshielded twisted pair or shielded twisted pair to provide 100-Mbps data communications.

CEPT   Conference of European Postal & Telecommunication Administration. The CEPT defines the E-1 signal format of 32 voice channels.

channel   A communication path that is normally fully duplex.

cladding   A low index of refraction layer of glass or other material that surrounds the fiber core, causing the light to stay captive in the core.

cleaver   An instrument used to cut optical fibers in such a way that the ends can be connected with low loss.

cluster controller   IBM models 3174, 3274, etc., control unit servicing several multiple IBM 3178, 3278, and other terminals.

CO   Central office. A local telephone company office that contains a switch terminating subscribers.

coating   A thin layer of plastic, or other material, usually 250 or 500 μm in diameter and color coded, covering the cladding of a fiber. Most fibers have a coating. It must be stripped for cleaving and splicing.

codec   A device that converts analog signals to digital signals. It also converts digital signals to analog signals.

concentrator   An electronic device used in LANs that allows a number of stations to be connected to a single data trunk.

conductor   A material allowing the flow of an electric current.

connector (fiber optic)   A device that joins two optical fibers together in a repeatable, low-optical-loss manner.

cord   A short length of flexible cable used for connecting equipment.

core   The inner portion of the fiber that carries light. Light stays in the core due to the difference in refractive index between the core and cladding.

coupler (fiber optic)   A device that joins three or more optical fiber ends together so that an optical signal can be split, or transmitted, from one fiber to two or more fibers.

CRC   Cyclic redundancy check. A mathematical checking method for determining the integrity of a data packet.

critical angle   The smallest angle at which a meridianal light ray can be totally reflected in a fiber core.

CSA   Canadian Standards Association.

CSMA/CD   Carrier sense multiple access with collision detection. An access protocol used by Ethernet/802.3.

CSU   Channel service unit. A device that allows testing functions on a channel such as loop back.

Cut-off wavelength   The shortest wavelength in which the fiber will propagate. Primarily in single-mode fibers.

dark current   The current that flows in a photo detector when there is no light on the detector.

dark fiber   Optical fibers in a cable that have no lightwave equipment connected to them (they are not lit by an optical source, hence dark).

data bandwidth   The maximum data rate that can be accommodated by a channel.

data rate   The amount of bits of information that can be transmitted per second. Expressed as Gbps, Mbps, kbps, or bps.

dB   Decibel. Measurement of optical power is 10 log (output power/input power).

dBm   Decibel, relative to 1 milliwatt. dBm = 10 log (output power milliwatts/1 milliwatt).

DCE   Data communication equipment. Communication equipment that is used as the interface to a communications channel for data terminal equipment (DTE) such as a modem.

dense WDM (DWDM)   A WDM that couples wavelengths, with channel spacing of 200 GHz or less, into a fiber.

dielectric   A material that will not conduct electricity under normal operating conditions (insulator).

differential group delay (DGD)   Accumulated delay of the two optical orthogonal polarization states due to polarization mode dispersion.

Digital   A data waveform format that has only two physical levels corresponding to 0s and 1s. *Also see* analog.

dispersion   Distortion of light signals in a fiber caused by fiber propagation characteristics.

dispersion compensating module (DCM)   A passive unit that can be used to counter dispersion effects in a fiber.

DMD   Differential mode delay effect.

DNA   Digital network architecture.

double window fiber   A fiber designed to operate at two different wavelengths.

DS0   Digital signal level 0. A digital communication channel that is 64 kbps. Twenty-four DS0s make one DS1 channel (T1).

DS1   Digital signal level 1. A digital communication channel that is 1.544 Mbps, T1 rate.

DS2   Digital signal level 2. A digital communication channel that is 6.312 Mbps, T2 rate.

DS3   Digital signal level 3. A digital communication channel that is 44.736 Mbps. It consists of 28 DS1s.

DSU   Digital service unit. A physical layer device that provides digital channel testing and monitoring functions.

DTE   Data terminal equipment. User terminal equipment such as a computer, terminal, or workstation.

DTMF   Dual tone multiple frequency. Also known as Touch-Tone.™ A set of audio frequencies used in telephone signaling. Often used by telephone sets.

E1   A European data communication standard rate of 2.048 Mbps, which can carry thirty 64-kbps channels.

E3   A European data communication standard rate of 34.368 Mbps, which can carry 16 E1 channels.

EIA   Electronic Industries Association.

EMI   Electromagnetic interference.

erbium-doped fiber amplifier (EDFA)   An optical amplifier that can amplify an optical signal without converting it into an electrical signal.

Ethernet   An 802.3 LAN technology that uses the CSMA/CD access method and a bus topology.

FBG   Fiber bragg gratings.

FDDI   Fiber distributed data interface. This is a fiber optic networking technology particularly suited for high-data-rate traffic. It uses a token-passing method and a dual counter rotating ring topology at a data rate of 100 Mbps.

ferrule   The rigid center portion of a fiber optic connector, usually steel or ceramic.

fiber bandwidth   The transmission frequency or wavelength which the signal magnitude decreases to half of its optical power (-3 dB).

fiber count   The number of optical fibers between any two locations. It can also refer to the number of optical fibers in a single cable.

fiber optics   The transmission of light through optical fibers for communication and signaling.

flame test rating (Ft 1, 4, 6)   A Canadian flame test rating for cables as per CSA C22.2 No. 0.3-M1985.

FOT   Fiber optic terminal.

frequency   The number of times a periodic action occurs in one second. Unit is cycles or hertz per second (Hz).

Fresnel reflection   Reflection of a portion of light at a surface between two materials having different indexes of refraction.

FT1   Fractional T1. A fraction of the full 24 channels in a T1 carrier.

FTTC   Fiber to the curb. The fiber optic architecture where fiber optics is deployed most of the way to the customer's home, but stops at the curb. It is terminated at the curb and uses copper pairs or coax to bring the signal into the home.

FTTH   Fiber to the home. The fiber optic architecture where fiber optics is deployed into the customer's home.

full duplex   The ability to transmit and receive signals at the same time.

full width half maximum (FWHM)   FWHM is the spectral width in nanometers (nm) of an optical source at one-half peak optical power level.

gainer (splice)   A splice gainer is an OTDR trace event that occurs at a fiber splice. It is shown by the OTDR as a rise in the trace instead of a drop at the splice. This occurs because of the increase in Rayleigh backscatter of the second fiber, caused by splicing two different fibers. An accurate splice reading can be made at this event by OTDR testing the event from both ends of the fiber, and then averaging the result.

giga   A prefix meaning one billion (10^9).

graded index fiber   An optical fiber in which the index of refraction of the core gradually decreases toward the cladding.

ground   A common electrical current return point to the earth usually through a ground rod.

ground loop currents   Undesirable ground currents that cause interference. Usually created when grounds are connected at more than one point.

half duplex   A communication method in which one end must wait until a transmission is complete in order to send information, and vice versa. Transmit and receive communications must alternate and cannot occur at the same time.

HDTV   High-definition television.

hertz (Hz)   One cycle per second.

hub   A communication device which uses a star wiring pattern topology common in LANs.

hybrid fiber/coax (HFC)   A cable that contains both optical fiber and coax cable.

ICEA   Insulated Cable Engineers Association, Inc.

IEEE   Institute of Electrical and Electronics Engineers.

impedance   The total opposition an electric circuit offers to an alternating current flow that includes both resistance and reactance.

index matching fluid   A liquid or gel with a refractive index that matches the core of the fiber.

index of refraction   The ratio of the speed of light in a vacuum to the speed of light in a material. Air $n = 1.003$; glass $n = 1.4$ to 1.6. The n also varies slightly for different wavelengths.

insertion loss   The optical power loss caused by insertion of an optical component such as a connector, splice, or attenuator.

insulator   A material that does not conduct electricity.

interbuilding cabling   Cabling between buildings.

intrabuilding cabling   Cabling within a building.

IRU   Indefeasible rights of use. An agreement for the right to use dark fiber bandwidth as stipulated in the agreement.

ISO   International Organization for Standards.

ITU   International Telecommunications Union.

ITU-T (formerly CCITT)   International Telecommunications Union—Telecommunications.

jacket   The outer coating of a wire or cable.

kilo   A prefix meaning one thousand (1000).

LAN   Local area network. A network in a local area, such as in one large building.

Laser   An acronym standing for "light amplification by stimulated emission of Radiation." A device that produces light, by the stimulated emission method, with a narrow range of wavelengths and emitted in a directional coherent beam. Fiber optic lasers are solid-state devices.

laser spectral bandwidth   The optical bandwidth occupied by a laser.

LED   An acronym for light-emitting diode. A semiconductor device that produces incoherent light when current passes through it.

light, fiber optic   The spectrum of light at 850-nm, 1300-nm, and 1500-nm wavelengths. These wavelengths are not visible to the human eye.

lightwave equipment   Any electronic communication equipment that is used for optical fiber transmission. It is also known as optical terminating equipment or optical modem.

LTE   Line-terminating equipment.

M13   Multiplexer DS1 to DS3.

MAN   Metropolitan area network. A network connecting multiple sites in one geographical area.

maximum coupling angle   The angle by which all incident light is totally and internally reflected by the optical fiber core. Maximum coupling angle = sin NA.

MC   Main cross connect.

mega   A prefix meaning one million (10^6).

micro   A prefix meaning one millionth (10^{-6}).

microbending loss   Loss in a fiber, caused by sharp curves of the core with displacements of a few micrometers. Such bends may be caused by the buffer, jacket, packaging, installation, etc. Losses can be significant over a distance.

milli   A prefix meaning one thousandth (10^{-3}).

minimum bend radius   The minimum radius of a curve that a fiber optic cable or optical fiber can be bent without any adverse effects to the cable or optical fiber characteristics.

mode   Refers to a single electromagnetic light wave that satisfies Maxwell's equations and the boundary conditions given by the fiber. It can simply be considered as a light ray path in a fiber.

mode field diameter (MFD)   The diameter of the light spot from a single-mode fiber. It is usually not the same diameter as the core diameter.

mode scrambler   A device that causes modes to mix in a fiber.

modem   An electronic device that converts one form of signal to another form using a modulation technique. An optical modem converts the electrical signal to an optical signal and vice versa.

modulation   Altering a carrier wave to carry signal information. Optical modulation involves altering the light wave amplitude or frequency to convey signal information.

MSDS   Material safety data sheet.

multimode fiber   A fiber that propagates more than one mode of light.

multiplexer   An electronic unit that combines two or more communication channels onto one aggregate channel.

nano   A prefix meaning one billionth (10^{-9}).

narrow-band WDM   A WDM (wavelength division multiplexer—see under W) that couples different wavelengths in the 1550-nm band onto a fiber.

NEC   National Electrical Code.

NEMA   National Electrical Manufacturers Association.

non-return to zero (NRZ)   A digital code where the signal level is too high for a 1 bit, too low for a 0 bit, and does not return too low for successive 1 bits.

numerical aperture   The sine of the angle measured between an incident light ray and the boundary axis, in which an optical fiber can accept and propagate light rays. It is a measure of the light-accepting property of the optical fiber.

NZ-DSF   Near-zero dispersion shifted fiber.

OCWR   Optical continuous wave reflectometer.

OFN   UL listing meaning nonconductive fiber optic cable. Each fiber cable bears the appropriate UL listing as required by the NEC. Additional limitations on fiber cable are covered in the *1992 National Electrical Code Handbook*.

OFNP   UL listing meaning nonconductive fiber optic cable plenums. Each fiber cable bears the appropriate UL listing as required by the NEC. Additional limitations on fiber cable are covered in the *1992 National Electrical Code Handbook*.

OFNR   UL listing meaning nonconductive fiber optic cable riser. Each fiber cable bears the appropriate UL listing as required by the NEC. Additional limitations on fiber cable are covered in the *1992 National Electrical Code Handbook*.

ohm   The electrical unit of resistance.

OPGW   Optical ground wire.

optical add/drop module (OADM)   A unit used to add or drop selected wavelengths in a fiber.

optical dynamic range   The receiver optical dynamic range is the light-level window in dBm at which a receiver can accept optical power.

optical fiber   A single optical transmission element comprised of a core, cladding, and coating. Commonly made of silica glass but can also be made from plastic.

optical loss   The reduction of fiber optic power caused by a substance.

optical margin   The value in dB of the difference between the total optical link loss and the manufacturer's equipment optical sensitivity.

optical return loss (ORL)   The total optical power reflected back to the input end of a fiber and represented in dB.

optical signal-to-noise ratio (OSNR)   The ratio of optical signal power to noise power and represented in dB.

optical time domain reflectometer (OTDR)   An optical time domain reflectometer is a test instrument that sends short light pulses down an optical fiber to determine the fiber's characteristics, attenuation, and length.

packet   A grouping of data with an address header and control information.

pad (optical)   A fixed optical attenuator.

patch cord (jumper), optical   A short length of buffered optical fiber, 3 mm in diameter, with connectors at both ends. It can be used to connect equipment to patch panels or jumper patch panels.

patch panel   A fiber cable termination panel.

PBX   Private branch exchange.

photodetector   A device that converts light energy into electrical energy. A silicon photo diode is commonly used in fiber optics.

pico   A prefix meaning 10^{-12}.

pigtail   A short length of buffered optical fiber with a connector at one end. It is used to terminate fibers.

PIN   Positive intrinsic negative photodiode.

PLC   Programmable logic controller.

plenum   An air space that can be found above dropped ceilings or in raised floors.

polarization dependent loss (PDL)   PDL is the loss in dB due to polarization mode dispersion.

polarization mode dispersion (PMD)   A single-mode fiber term. Represents the dispersion of optical light caused by two states of polarized light traveling at slightly different velocities and thereby arriving at their destination at different times. It is measured in units $ps/ps/\sqrt{km}$.

polyethylene   A thermoplastic material often used for cable jackets.

polyvinyl chloride   A thermoplastic material often used for cable jackets.

pop Point of presence. A physical location where a carrier provides services to a customer.

POTS Plain old telephone system. A two-wire loop start ring telephone connection with Touch Tone or make or break signaling.

protocol A set of rules that will enable data communications.

receiver, optical An electronic unit that converts light signals to electrical signals.

reflectance Reflectance is measured in dB and represents the reflected optical power from a single event in a fiber.

repeater A device that repeats and regenerates a signal.

return to zero (RZ) A digital code where the signal level is high for a 1 bit for the first half of the bit interval and then goes to low for the second half of the bit interval. The level stays low for a 0 bit complete interval.

RFI Radio frequency interference.

router A LAN device that operates at the OSI protocol stack network layer and is used to connect dissimilar LANs or networks.

SDH Synchronous digital hierachy.

sensitivity The minimum amount of optical power needed to be received by the lightwave equipment to achieve signal transmission to equipment specification.

single-mode fiber A fiber that carries only one mode of light. Only one light path is propagated.

SNA Systems network architecture. IBM's seven-layer data communication architecture.

SONET Synchronous optical network. A protocol standard for communication over optical fiber defined in layer 1 (physical layer) of the OSI protocol stack.

spectral bandwidth A group of optical wavelengths (or frequencies) occupied by or reserved for optical sources. Not to be confused with data bandwidth.

speed of light 2.998×10^{8} meters per second in a complete vacuum. It is less in any other material.

splice A permanent junction between two fibers. It can be made by using a fusion splice method or a mechanical splice method.

step index fiber An optical fiber in which the core has a constant index of refraction.

STP Shielded twisted-pair wire.

stratum A primary reference clock used for network synchronization.

STS-1 Synchronous transport signal level 1.

synchronous signal A signal that is synchronized to a network clock.

synchronous transmission   Data communication protocol where the data is sent continuously.

T1   A communications aggregate of a data rate of 1.544 Mbps that has 24 channels.

transmitter, optical   An electronic unit that converts electrical signals to light signals.

UL   Underwriters' Laboratories.

UTP   Unshielded twisted pair wire.

VCSEL   Vertical cavity-surface emitting laser.

VOA   Variable optical attenuator. A device that is used to introduce attenuation into a fiber span. Also known as a pVOA, which is a programmable VOA. The pVOA attenuation is electrically controlled and can be set from a remote location.

WAN   Wide area network. A network that extends into different cities.

wavelength division multiplexer (WDM)   A unit that can couple and decouple more than one optical wavelength (λ) onto one optical fiber.

wavelength division multiplexing   The combining of two or more optical signals of different wavelengths onto one fiber.

wide-band WDM   A WDM that couples 1310-nm and 1550-nm wavelengths onto a fiber.

APPENDIX B
UNITS

Physical constants

Speed of light in vacuum:	$c = 2.998 \times 10^8$ m/s
Planck's constant:	$h = 6.63 \times 10^{34}$ J·s
Boltzmann's constant:	$k = 1.38 \times 10^{-23}$ J/K
Electron charge:	$e = -1.6 \times 10^{-19}$ C

Prefixes used to indicate multiples

Prefix	Factor	Symbol
exa-	10^{18}	E
peta-	10^{15}	P
tera-	10^{12}	T
giga-	10^{9}	G
mega-	10^{6}	M
kilo-	10^{3}	k
hecto-	10^{2}	h
deca-	10	da
deci-	10^{-1}	d
centi-	10^{-2}	c
milli-	10^{-3}	m
micro-	10^{-6}	μ
nano-	10^{-9}	n
pico-	10^{-12}	p
femto-	10^{-15}	f
atto-	10^{-18}	a

Units

Unit	Symbol	Measurement
meter	m	length
gram	g	mass
volt	V	voltage
ohm	Ω	resistance
ampere	a	current
coulomb	C	charge
watt	w	power
joule	J	energy
farad	F	capacitance
second	s	time
celsius	C	temperature
kelvin	K	temperature
bit	b	pulse
byte	byte	8 bits

Digital transmission rates

Designation	Data Rate	Voice Channels (64 kbps)
DS0	64 kbps	1
FT1	(1 to 24)×64 kbps	1 to 24
T1 - DS1	1.544 Mbps	24
T2 - DS2	6.312 Mbps	96
T3 - DS3	44.736 Mbps	672
T4 - DS4	274.175 Mbps	4,032

SONET/SDH transmission rates

SONET	SDH	Data Rate	Voice Channels (64 kbps)
OC-3	STM-1	155.52 Mbps	2,016
OC-12	STM-4	622.08 Mbps	8,064
OC-48	STM-16	2488.32 Mbps	32,256
OC-192	STM-64	9.953 Gbps (10 Gbps)	129,024
OC-768*	STM-256	39.813 Gbps (40 Gbps)	516,096

*Note: At the time of writing this book, OC-768/STM-256 systems have not yet been introduced into the market, but a number of manufacturers are developing them in the lab.

APPENDIX C
OPTICAL FIBER COLOR CODES

	Common	
Fiber Number	**Tube Color**	**Fiber Color**
1	Blue	Blue
2	Blue	Orange
3	Blue	Green
4	Blue	Brown
5	Blue	Grey
6	Blue	White
7	Blue	Red
8	Blue	Black
9	Blue	Yellow
10	Blue	Violet
11	Blue	Rose
12	Blue	Aqua
13	Orange	Blue
14	Orange	Orange
15	Orange	Green
16	Orange	Brown
17	Orange	Grey
18	Oange	White
19	Orange	Red
20	Orange	Black
21	Orange	Yellow
22	Orange	Violet
23	Orange	Rose
24	Orange	Aqua
25	Green	Blue
26	Green	Orange
27	Green	Green
28	Green	Brown
29	Green	Grey
30	Green	White
31	Green	Red
32	Green	Black

Other Common Color Codes:		
Cable Manufacturer A—Possible Alternate		
Fiber Number	**Tube Color**	**Fiber Color**
1	Black	Blue
2	Black	Orange
3	Black	Green
4	Black	Brown
5	Black	Grey
6	Black	White
7	Red	Blue
8	Red	Orange
9	Red	Green
10	Red	Brown
11	Red	Grey
12	Red	White
13	Yellow	Blue
14	Yellow	Orange
15	Yellow	Green
16	Yellow	Brown
17	Yellow	Grey
18	Yellow	White
19	Violet	Blue
20	Violet	Orange
21	Violet	Green
22	Violet	Brown
23	Violet	Grey
24	Violet	White

Cable Manufacturer B—Possible Alternate		
Fiber Number	**Tube Color**	**Fiber Color**
1	Red	Brown
2	Red	Blue
3	Red	Orange
4	Red	White
5	Red	Grey
6	Red	Green
7	Black	Brown
8	Black	Blue
9	Black	Orange
10	Black	White
11	Black	Grey
12	Black	Green
13	Yellow	Brown
14	Yellow	Blue
15	Yellow	Orange
16	Yellow	White
17	Yellow	Grey
18	Yellow	Green
19	Green	Brown
20	Green	Blue
21	Green	Orange
22	Green	White
23	Green	Grey
24	Green	Green

Cable Manufacturer C—Possible Alternate		
Fiber Number	**Tube Color**	**Fiber Color**
1	Blue	Blue
2	Blue	Orange
3	Blue	Green
4	Blue	Brown
5	Blue	Grey
6	Blue	White
7	Orange	Blue
8	Orange	Orange
9	Orange	Green
10	Orange	Brown
11	Orange	Grey
12	Orange	White
13	Green	Blue
14	Green	Orange
15	Green	Green
16	Green	Brown
17	Green	Grey
18	Green	White

APPENDIX D
FIBER OPTIC RECORDS

Optical Fiber Attenuation/Loss Test Results

Date_____   Tested by_____

Cable Location_____   Cable Length (OTDR)_____(Physical)_____

Start Cable Jacket Sequence No._____   End Cable Jacket Sequence No._____

OTDR Location_____   Fiber Type_____

Index of Ref._____   Cable Manufacturer_____

OTDR Type_____   Power Meter/Source Type_____

Fiber Number and Bundle Color/Fiber Color	Test Wavelength_____nm			Test Wavelength_____nm		
	OTDR Measurements		Power Meter Loss dB	OTDR Measurements		Power Meter Loss dB
	Average Attenuation dB/km	End to End Loss dB		Average Attenuation dB/km	End to End Loss dB	
1.						
2.						
3.						
4.						
5.						
6.						
7.						
8.						
9.						
10.						
11.						
12.						
13.						
14.						
15.						
16.						
17.						
18.						
19.						
20.						
21.						
22.						

Optical Splice and Anomaly Test Results

Date_____ Tested by_____

Cable Location_____ Cable Length (OTDR)_____(Physical)_____(Factor)_____

Start Cable Jacket Sequence No._____ End Cable Jacket Sequence No._____

Location A_____ Fiber Type_____

Location B_____ Cable Manufacturer_____

OTDR Type_____

Fiber Number _____	Test Wavelength_____nm			Test Wavelength_____nm		
	OTDR Measurements			OTDR Measurements		
Splice or Anomaly Location from A in km	From Location A Loss dB	From Location B Loss dB	Average Loss (A + B)/2 dB	From Location A Loss dB	From Location B Loss dB	Average Loss (A + B)/2 dB

Equipment Optical Power Output Record

Date_____ Tested By_____

Location _____

Power Meter Model_____

Equipment Type	Location	Fiber Jumper FPP Position	Output Power dBm

Optical Fiber Return Loss Record

Date_____   Tested By_____

Location _____

Return Power Loss Meter Model_____

Fiber Panel Location	Fiber FPP Number	Return Loss dB

Optical Link Budget Estimate

Optical Fiber Diameter: Core _____ μm / Cladding _____ μm

Optical Fiber Numerical Aperture (NA): _____

Lightwave Equipment Operating Wavelength: _____ nm

	Description	
a.	Optical Fiber Loss at Operating Wavelength:_____ _____ km length at _____ dB / km	_____ dB
b.	Splice Loss: _____splices x _____dB / splice	_____ dB
c.	Connection Loss: _____connections x _____ dB / connection	_____ dB
d.	Other Component Losses:	_____ dB
e.	Design Margin:	_____ dB
f.	**Total Link Loss** (a + b + c + d + e)	**_____ dB**
g.	Transmitter Average Output Power:	_____ dBm
h.	**Receiver Input Power** (g - f)	**_____ dBm**
i.	Receiver Dynamic Range:	_____ dB
j.	Receiver Sensitivity for (BER or S/N)	_____ dBm
k.	**Remaining Margin:** (h - j)	**_____ dB**

Notes:

1. Remaining Margin, line k, must be greater than zero for proper system design.

2. In line c, exclude lightwave equipment optical fiber connection loss.

3. Receiver Input Power, line h, must be within Receiver Dynamic Range, line i, to prevent receiver optical saturation.

APPENDIX E
RELEVANT STANDARDS AND REFERENCES

TIA/EIA 568A	Commercial Building Telecommunications Cabling Standard
TSB 72	Centralized Optical Fiber Cabling Guidelines, Supplement to TIA/EIA 568A
TIA/EIA-455-34A	Interconnection Device Insertion Loss Test
TIA/EIA-455-107	Determination of Component Reflectance or Link/System Return Loss Using a Loss Test Set.
TIA/EIA-455-171	Attenuation by Substitution Measurement for Short-Length Multimode Graded Index and Single-Mode Optical Fiber Cable Assemblies
TIA/EIA-359A	Standard Colors for the Color Identification and Coding
TIA/EIA-455-3A	Procedure to Measure Temperature Cycling Effects on Optical Fibers, Optical Cable, and Other Passive Fiber Optic Components
TIA/EIA-455.25A	Repeat Impact Testing of Fiber Optic Cables and Cable Assemblies
TIA/EIA-455-31A	Fiber Tensile Proof Method
TIA/EIA-455-41	Compressive Loading Resistance of Fiber Optic Cables
TIA/EIA-455-45A	Microscopic Method for Measuring Fiber Geometry of Optical Waveguide Fibers
TIA/EIA-455-55A	Method for Measuring the Coating Geometry of Optical Fibers
TIA/EIA-455-59	Use of OTDR for Fiber Optic Point Defects

TIA/EIA-455-62	Optical Fiber Macrobend Attenuation
TIA/EIA-455-78A	Spectral Attenuation Cutback Measurement for Single-Mode Optical Fiber
TIA/EIA-455-81	Compound Flow (Drip) Test for Filled Optic Fiber Cable
TIA/EIA-455-82A	Fluid Penetration Test for Fluid Blocked Fiber Optic Cable
TIA/EIA-455-85	Fiber Optic Cable Twist Test
TIA/EIA-455-104	Fiber Optic Cable Cyclic Flexing Test
TIA/EIA-455-164	Single-Mode Fiber Measurement of Mode Field Diameter by Far-Field Scanning
TIA/EIA-455-167	Mode Fields Diameter Measurement—Variable Aperture Method in the Far Fields
TIA/EIA-455-168	Chromatic Dispersion Measurement of Multimode Graded Index and Single-Mode Optical Fibers by Spectral Group Delay Measurement in the Time Domain
TIA/EIA-455-170	Measurement Method for Optical Fiber Geometry by Automated Grey-Scale Analysis
TIA/EIA-455-174	Mode Field Diameter of Single-Mode Optical Fiber by Knife-Edge Scanning in the Far Field
TIA/EIA-455-175	Chromatic Dispersion Measurement of Optical Fibers by the Differential Phase Shift Method
TIA/EIA-455-176	Measurement Method of Optical Fiber Geometry by Automated Grey-Scale Analysis

APPENDIX F
DWDM 50-GHZ CHANNEL SPACING

Wavelength, nm	Frequency, THz	Wavelength, nm	Frequency, THz	Wavelength, nm	Frequency, THz	Wavelength, nm	Frequency, THz
1525.66	196.50	1535.43	195.25	1545.32	194.00	1555.34	192.75
1526.05	196.45	1535.82	195.20	1545.72	193.95	1555.75	192.70
1526.44	196.40	1536.22	195.15	1546.12	193.90	1556.15	192.65
1526.83	196.35	1536.61	195.10	1546.52	193.85	1556.55	192.60
1527.21	196.30	1537.00	195.05	1546.92	193.80	1556.96	192.55
1527.60	196.25	1537.40	195.00	1547.32	193.75	1557.36	192.50
1527.99	196.20	1537.79	194.95	1547.72	193.70	1557.77	192.45
1528.38	196.15	1538.19	194.90	1548.11	193.65	1558.17	192.40
1528.77	196.10	1538.58	194.85	1548.51	193.60	1558.58	192.35
1529.16	196.05	1538.98	194.80	1548.91	193.55	1558.98	192.30
1529.55	196.00	1539.37	194.75	1549.32	193.50	1559.39	192.25
1529.94	195.95	1539.77	194.70	1549.72	193.45	1559.79	192.20
1530.33	195.90	1540.16	194.65	1550.12	193.40	1560.20	192.15
1530.72	195.85	1540.56	194.60	1550.52	193.35	1560.61	192.10
1531.12	195.80	1540.95	194.55	1550.92	193.30	1561.01	192.05
1531.51	195.75	1541.35	194.50	1551.32	193.25	1561.42	192.00
1531.90	195.70	1541.75	194.45	1551.72	193.20	1561.83	191.95
1532.29	195.65	1542.14	194.40	1552.12	193.15	1562.23	191.90
1532.68	195.60	1542.54	194.35	1552.52	193.10	1562.64	191.85
1533.07	195.55	1542.94	194.30	1552.93	193.05	1563.05	191.80
1533.47	195.50	1543.33	194.25	1553.33	193.00	1563.45	191.75
1533.86	195.45	1543.73	194.20	1553.73	192.95	1563.86	191.70
1534.25	195.40	1544.13	194.15	1554.13	192.90	1564.27	191.65
1534.64	195.35	1544.53	194.10	1554.54	192.85	1564.68	191.60
1535.04	195.30	1544.92	194.05	1554.94	192.80		

APPENDIX G
SIGNAL GROUNDING CONDUCTOR SIZE

As per standard IEEE 241-1990 "Grey Book," at least 2000 circular mils of cross-sectional area per linear foot of ground conductor.

Conductor Size (AWG)	Maximum Area (CM)	Conductor Length (feet)
6	26,240	13
4	41,740	21
3	52,620	26
2	66,360	33
1	83,690	42
1/0	105,600	53
2/0	133,100	67
3/0	167,800	84
4/0	211,600	106
250 MCM	250,000	125
300 MCM	300,000	150
350 MCM	350,000	175
400 MCM	400,000	200
500 MCM	500,000	250
600 MCM	600,000	300
700 MCM	700,000	350
800 MCM	800,000	400
900 MCM	900,000	450
1000 MCM	1,000,000	500
1250 MCM	1,250,000	625
1500 MCM	1,500,000	750
1750 MCM	1,750,000	875
2000 MCM	2,000,000	1000

Follow TIA/EIA-607 for grounding bus bar size design. Refer to the NEC book and any local standards for proper electrical power grounding standards.

APPENDIX H
WAVELENGTH DIVISION MULTIPLEXERS (WDMs)

WHAT IS A WDM (WAVELENGTH DIVISION MULTIPLEXER)?

* A fiber optic device that couples two or more optical wavelengths from separate fibers into one fiber, and vice versa (Figs. H.1 and H.2).

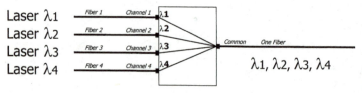

FIGURE H.1   Four-channel WDM.

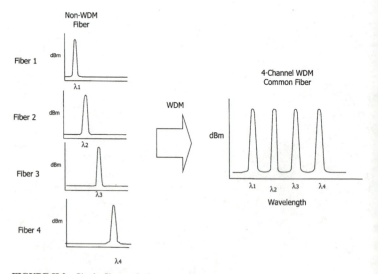

FIGURE H.2   Single-fiber optical spectrum.

WHAT DOES A WDM LOOK LIKE?

Most commonly available as a small cassette:

- With fiber adapters, to be used with fiber jumpers
- With fiber pigtails, for splicing directly to fiber cable (Figs. H.3 and H.4).
- Completely passive, no power connection required.

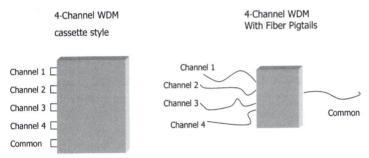

FIGURE H.3   What does a WDM look like?

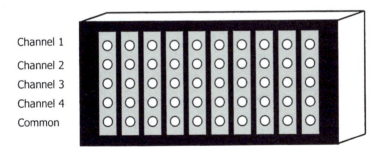

10 WDMs mounted in one 19- or 23-inch shelf

FIGURE H.4   WDM shelf.

WHY USE WDMs?

Advantages

1. Can reduce the number of fibers required for a communication system by increasing the available capacity of a fiber, more systems can be connected using one fiber cable, can reduce OSP construction, and enables significant cost savings $$$. (Figs. H.5 and H.6)

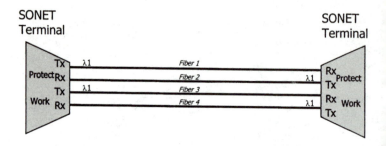

4 fibers required for a SONET linear system

FIGURE H.5   SONET without WDMs.

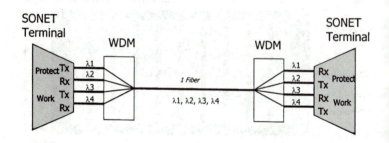

1 fiber possible for a SONET linear system using WDMs

FIGURE H.6   SONET with WDMs.

2. May be possible to retrofit existing fiber optic cable networks with WDMs to increase fiber cable capacity, reducing OSP fiber construction, and enabling significant cost savings.

3. Leasing wavelengths (λ) instead of fibers can bring in additional revenue.

4. Amplification of optical signals from numerous systems made possible by deployment of one optical amplifier at an intermediate node. Alternative to the traditional method of installing a regenerator for each individual system (Fig. H.7).

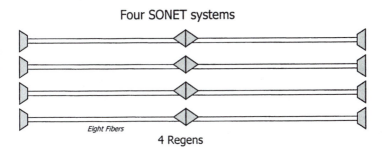

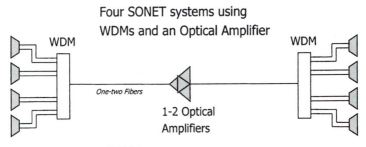

FIGURE H.7 Why use WDMs?

5. WDMs can multiplex different optical communication protocols onto one fiber (Fig. H.8):

- SONET
- Fiber Channel
- IP over fiber
- ATM over fiber
- Others

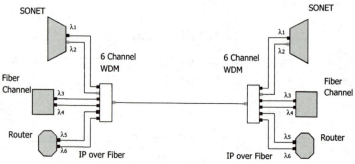

FIGURE H.8 Why use WDMs?

Disadvantages

1. Need to use lasers with different wavelengths, tuned to each WDM channel.

2. Adds optical loss to a fiber link.

3. Adds complexity to a fiber system.

4. More difficult to provision and test.

HOW ARE WDMs USED?

- WDMs are used in pairs, one at each end of an optical fiber.
- WDM pairs need to be matched:
  - Type, mux and demux, cross-band, narrow-band, dense, bidirectional, unidirectioal
  - Number of wavelengths
  - Wavelength spacing (Fig. H.9)

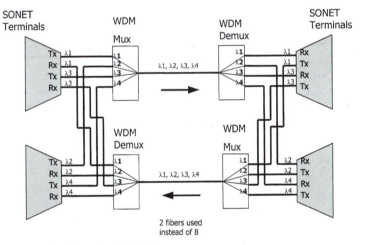

FIGURE H.9   Narrow-band WDM pair.

WHERE ARE WDMs PLACED?

- Inserted between the outside plant fiber distribution panel and lightwave equipment.
- Connected to lightwave equipment and fiber distribution panel with fiber jumpers (Fig. H-10).

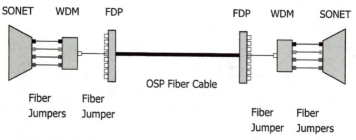

FIGURE H.10   Where are WDMs placed?

WHAT ARE THE DIFFERENT TYPES OF WDMs?

- Wide-band/cross-band (Fig. H.11)
- Narrow-band (Fig. H.12)
- Dense (ultra dense) (Fig. H.13)
- Unidirectional (Fig. H.14)
- Bidirectional (Fig. H.15)
- Universal (Fig. H.16)
- OADM (optical add drop mux)

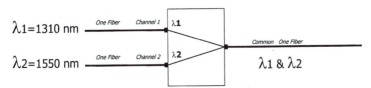

FIGURE H.11   Wide-band/cross-band WDM.

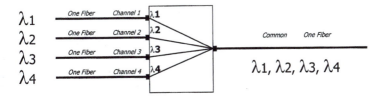

$\lambda 1, \lambda 2, \lambda 3, \lambda 4$ in 1550 nm Band
Channel Spacing 1000 GHz (8 nm)

FIGURE H.12   Narrow-band WDM.

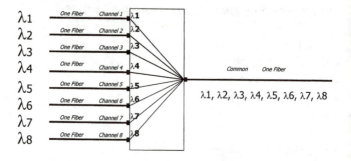

$\lambda1, \lambda2, \lambda3, \lambda4, \lambda5, \lambda6, \lambda7, \lambda8... \lambda64$ in 1550 nm Band

Channel Spacing of 100 GHz (0.8 nm) or 50 GHz (0.4 nm) ITU-T Standard

FIGURE H.13   Dense WDM.

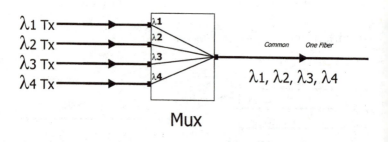

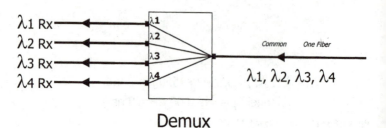

FIGURE H.14   Unidirectional WDM.

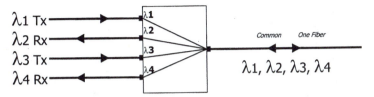

FIGURE H.15   Bidirectional WDM.

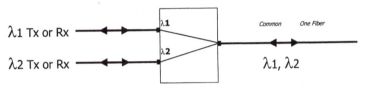

FIGURE H.16   Universal WDM.

OPTICAL ADD/DROP MULTIPLEXER (OADM)

- Type of WDM that allows one or more wavelengths to be inserted or dropped at a node (Fig. H.17). See also Figs. H.18 to H.23.

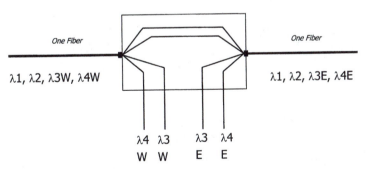

FIGURE H.17   Optical add/drop multiplexer (OADM).

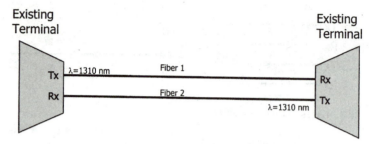

Existing 1310-nm
2-fiber communication
system.

FIGURE H.18   Simple wide-band WDM application which doubles fiber capacity.

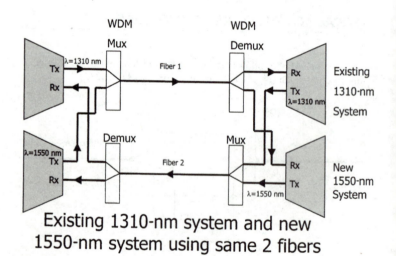

Existing 1310-nm system and new
1550-nm system using same 2 fibers

FIGURE H.19   Simple wide-band WDM application.

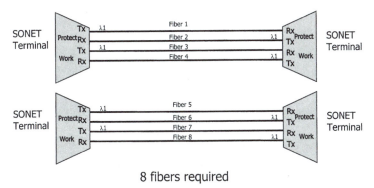

8 fibers required

FIGURE H.20   Narrow-band WDM application. Two SONET systems without WDMs.

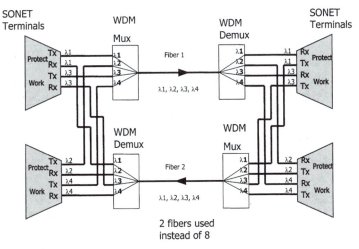

2 fibers used
instead of 8

FIGURE H.21   Narrow-band WDM application.

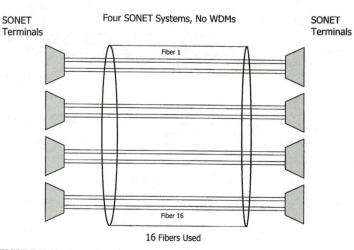

FIGURE H.22   DWDM application.

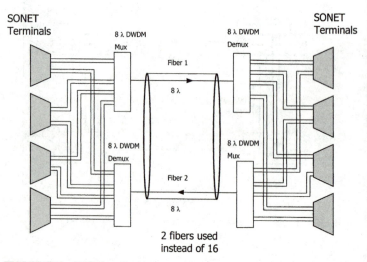

FIGURE H.23   DWDM application.

PRESENT WDM CAPABILITY

- Present commercial DWDM systems that can place 64 wavelengths onto one fiber are available.
- One fiber can connect 16 SONET OC-192s (10-Gbps systems).
- Resulting in 160 Gbps of communication on one fiber.

THE FUTURE OF WDMs

- An increase of over 1000 wavelengths on one fiber.
- Will allow 250 SONET OC-192s (10-Gbps systems) to be connected with one fiber.
- Will result in over 2.5 Tbps of communication in one fiber.

SUMMARY

- WDMs can reduce fiber network cost by increasing existing fiber cable capacity.
- With WDM deployment, it is essential that proper lasers are used.
- WDMs can work with different optical communication protocols.
- Before deploying new fiber optic communication systems or cables, assess your fiber network to determine if savings can be achieved by deploying WDMs.
- If WDMs are deployed, WDM training for maintenance/operations crews should be provided.

INDEX

ABOUT THE AUTHOR

Bob Chomycz, P. Eng., is founder and president of Telecom Engineering, Inc., a company that specializes in fiber optic telecommunication system engineering and construction. With over 15 years of experience as a telecom engineer, Bob has accumulated a wealth of knowledge in fiber optics, which he shares with you in this book. His experience includes engineering design, integration, installation, and commissioning a large variety of fiber optic projects for data, LAN, WAN, voice, and video communication, throughout buildings and across cities. He has also trained many groups in fiber optic theory and proper installation techniques. At present, he is involved in helping industry build state-of-the-art fiber optic communication systems. He is also the author of *Fiber Optic Installations*, published by McGraw-Hill in 1996.